IS SCIENCE GETTING
OUT OF HAND?

"The age of innocent faith in science and technology may be over. We were given a spectacular signal of this change on a night in November 1965. On that night all electric power in an 80,000-square mile area of the northeastern United States and Canada failed. The breakdown was a total surprise. For hours engineers and power officials were unable to turn the lights on again; for days no one could explain why they went out; even now no one can promise that it won't happen again."

(From the book)

SCIENCE AND SURVIVAL begins with the blackout—the symbol of technological breakdown—and proceeds to question our headlong rush to create the potential for even greater disaster. We court that disaster when we link everything to everything else and fail to consider the ultimate impact on ourselves and this planet.

Barry Commoner calls for us to heed this warning now while there is still time to correct—while there is still time to take the steps to avoid destruction.

SCIENCE AND SURVIVAL

Barry Commoner

BALLANTINE BOOKS • NEW YORK
An Intext Publisher

SBN 345-02084-7-125
This edition published by arrangement with
The Viking Press, Inc.

First Printing: November, 1970

Front cover designed concept, courtesy of Gary Friedman

Back cover photograph courtesy of Washington
University, St. Louis.

Printed in the United States of America

BALLANTINE BOOKS, INC.
101 Fifth Avenue, New York, N.Y. 10003

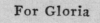
For Gloria

Acknowledgments

Most of the ideas expressed in this book were developed during the course of a number of activities, of varying degrees of formal organization, in which I have engaged in the last few years. Among these are the Committee on Science in the Promotion of Human Welfare of the American Association for the Advancement of Science, the St. Louis Committee for Nuclear Information (CNI), and the Scientists' Institute for Public Information. I owe a great deal to my colleagues and friends in these organizations, for many of my ideas about the interactions between science and society have emerged from numerous, deeply illuminating discussions with them. Several of them have been particularly helpful in the preparation of this book. Mrs. Virginia Brodine, editor of *Scientist and Citizen*, the CNI publication, has been a constant source of valuable advice in the preparation of this and related writings. I am grateful too for the work of Mr. Gorman L. Mattison and Mr. Sheldon Novick, who, in their capacity as administrative assistants to the AAAS Committee, as members of the CNI staff, and in innumerable less formal ways, have, by their effective analysis of specific problems, by their valuable insights, and through a deep commitment to the issue of applying science to human welfare, made important contributions to many of my own activities. I also wish to thank Mrs. Anabelle Sylvester and Mrs. Gladys Yandell for peerless secretarial work on this and related manuscripts.

Contents

SCIENCE AND SURVIVAL

1

Is Science Getting Out of Hand?

THE age of innocent faith in science and technology may be over. We were given a spectacular signal of this change on a night in November 1965. On that night all electric power in an 80,000-square-mile area of the northeastern United States and Canada failed. The breakdown was a total surprise. For hours engineers and power officials were unable to turn the lights on again; for days no one could explain why they went out; even now no one can promise that it won't happen again.

The failure knocked out a huge network which was supposed to shift electric power from areas with excess generating capacity to those facing a heavy drain. But on that night the power grid worked against its intended purpose. Instead of counteracting a local power failure, it spread the trouble out of control until the whole system was engulfed and dozens of cities were dark.

The trouble began with the failure of a relay which controlled the flow of electricity from the Sir Adam Beck No. 2 power plant in Queenston, Ontario, into one of its feeder lines. The remaining lines, unable to carry the extra load, shut down their own safety switches. With these normal exits blocked the plant's full power flowed back along

3

the lines that tied the Queenston generators into
the U.S.-Canadian grid. This sudden surge of
power, traveling across New England, quickly
tripped safety switches in a series of local power
plants, shutting them down. As a result the New
England region, which until then had been feeding excess electricity into the Consolidated Edison
system in New York, drained power away from
that city; under this strain the New York generators were quickly overloaded and their safety
switches shut off. The blackout was then complete. The system had been betrayed by the very
links that were intended to save local power plants
from failure.[1]

In one of the magazine reports of the great
blackout, there is a photograph[2] that tells the
story with beautiful simplicity. It shows a scene in
Consolidated Edison's Energy Control Center.
Stretched purposefully across the photograph is an
operational diagram of the New York power system; an intricate but neat network of connections,
meters, and indicators symbolizing the calculated
competence of this powerful machine. In the fore-
ground, dwarfed by the diagrammatic system and
in curious contrast to its firm and positive lines, is
a group of very puzzled engineers.

This same contrast between man and machine
is expressed in the accompanying text:

The Northeast grid was magnificently intercon-
nected and integrated. But only machines spoke
over it, one to the other. They asked each other
mechanical questions and gave each other mechanical responses. No human responsibility had immediate control over this entire system. Thus, no

human being can answer the still-unanswered question: Why?

But this electronic thinking did not protect the people of the city.

It was required that New York come to the brink of chaos to refresh an old truth: People—men of frailty, judgment and human decisions—must control machines. Not vice versa.[3]

One man, however, if he had lived to see it, would not have been surprised by the great blackout—Nobert Wiener, the mathematician who did so much to develop cybernetics, the science which guides the design of complex electrical grids and their computerized controls. Cybernetics has produced electronic brains and all the other marvelous machines that now operate everything from election reports to steel plants; that have made the robot no longer a cartoon but a reality; that made the U.S.-Canadian power grid feasible.

Just six years before the blackout Dr. Wiener reviewed a decade of remarkable progress in the science which he helped to create.[4] He reported at that time on the development of a new kind of automatic machine, a computer that had been programmed to play checkers. Engineers built into the electronic circuits a correct understanding of the rules of checkers and also a way of judging what moves were most likely to beat the computer's opponents. The computer made a record of its opponent's moves in the current and previous games. Then, at great speed, it calculated its opponent's most likely moves in any given situation and, having figured those out, adjusted its own game, move by move, to give itself the best chance of winning. The engineers designed a machine that not only knew how to play checkers but could learn from experience and actually improve its own game.

Dr. Wiener described the first results of the checkers tournaments between the computer and its programmers. The machine started out playing an accurate but uninspired game which was easy to beat. But after about ten or twenty hours of practice the machine got the hang of it, and from then on the human player usually lost and the machine won.

Dr. Wiener emphasized this point: Here was a machine designed by a man who built into it everything that it could do. Yet, because it could calculate complicated probabilities faster than the man could, the machine learned to play checkers against the man better than he could against the machine. Dr. Wiener concluded that it had become technically possible to build automatic machines that "most definitely escape from the complete effective control of the man who has made them."

The U.S.-Canadian power grid is just such a machine. By following the rules built into its design, the machine acted—before the engineers had time to understand and countermand it—in a way that went against their real wishes.

One month after the great blackout, there occurred in Salt Lake City, Utah, a little-noticed event that can take its place beside the power failure as a monument to the blunders which have begun to mar the accomplishments of modern science and technology. There, nine children from Washington County, Utah, entered a hospital for tests to determine whether abnormal nodules in their thyroid glands were an indication of possible thyroid disease: nontoxic goiter, inflammation, benign or malignant tumors. Fifteen years earlier these children had been exposed to radioactive iodine produced by fallout from the nearby Nevada atomic test site.

It will be some time before any one can tell whether the incidence of thyroid nodules in this group of children is statistically significant, and if so, whether the nodules are really due to fallout. But regardless of the outcome, the mere fact that health authorities felt compelled to look for an effect of fallout on the health of these children is itself a surprise.[5]

The chain of events which brought the children into the hospital began in the 1950s when the AEC started a long series of nuclear explosions at its Nevada test site in the conviction that " ... these explosives created no immediate or long-range hazard to human health outside the proving ground." But among the radioactive particles of the fallout clouds that occasionally escaped into the surrounding territory was the isotope iodine-131. As these clouds passed over the Utah pastures, iodine-131 was deposited on the grass; being widely spread, it caused no alarming readings on outdoor radiation meters. But dairy cows grazed these fields. As a result, iodine-131, generated in the mushroom cloud, drifted to Utah farms, was foraged by cows, passed to children in milk, and was gathered in high concentration in the children's thyroid glands. Here in a period of a few weeks the iodine-131 released its radiation. If sufficiently intense, such radiation passing through the thyroid cells may set off subtle changes which, though quiescent and hidden for years, eventually give rise to disease.

Like the Northeast blackout, this too is a chain reaction. Where the blackout reaction chain took minutes, the iodine-131 chain took days and in a sense years. But in both cases the process was over and the damage done before we understood what had happened.

Modern science, and the huge technological enterprises which it produces, represent the full flowering of man's understanding of nature. Scientific knowledge is our best guide to controlling natural forces. In this it has been magnificently successful; it is this success which has given us the marvels of modern electricity, and the tremendous power of nuclear bombs.

The power blackout and the Utah thyroid problem have cast a shadow—small, but deeply troubling—over the brilliance of these scientific successes. Is it possible that we do not know the full consequences of the new power grids and the new bombs? Are we really in control of the vast new powers that science has given us, or is there a danger that science is getting out of hand?

2

Sorcerer's Apprentice

WE are surrounded by the technological successes of science: space vehicles, nuclear power, new synthetic chemicals, medical advances that increase the length and usefulness of human life. But we also see some sharp contrasts. While one group of scientists studies ways to provide air for the first human visitors to the moon, another tries to learn why we are fouling the air that the rest of us must breathe on earth. We hear of masterful schemes for using nuclear explosions to extract pure water from the moon; but in some American cities the water that flows from the tap is undrinkable and the householder must buy drinking water in bottles. Science is triumphant with far-ranging success, but its triumph is somehow clouded by growing difficulties in providing for the simple necessities of human life on the earth.

Our Polluted Environment

For about a million years human beings have survived and proliferated on the earth by fitting unobtrusively into a life-sustaining environment, joining a vast community in which animals, plants, microorganisms, soil, water, and air are tied together in an elaborate network of mutual relation-

ships. In the preindustrial world the environment
appeared to hold an unlimited store of clean air
and water. It seemed reasonable, as the need
arose, to vent smoke into the sky and sewage into
rivers in the expectation that the huge reserves of
uncontaminated air and water would effectively
dilute and degrade the pollutants—perhaps in the
same optimistic spirit that leads us to embed
slotted boxes in bathroom walls to receive razor
blades. But there is simply not enough air and
water on the earth to absorb current man-made
wastes without effect. We have begun to merit the
truculent complaint against the works of the pale-
face voiced by Chief Satinpenny in Nathanael
West's A *Cool Million*: "Now even the Grand Can-
yon will no longer hold razor blades."

Fire, an ancient friend, has become a man-made
threat to the environment through the sheer quantity of the waste it produces. Each ton of wood,
coal, petroleum, or natural gas burned contributes
several tons of carbon dioxide to the earth's atmosphere. Between 1860 and 1960 the combustion of
fuels added nearly 14 percent to the carbon-
dioxide content of the air, which had until then
remained constant for many centuries. Carbon dioxide plays an important role in regulating the
temperature of the earth because of the "greenhouse effect." Both glass and carbon dioxide tend
to pass visible light but absorb infrared rays. This
explains why the sun so easily warms a greenhouse
on a winter day. Light from the sun enters
through the greenhouse glass. Within, it is absorbed by soil and plants and converted to infrared heat energy which remains trapped inside
the greenhouse because it cannot pass out again
through the glass. Carbon dioxide makes a huge

greenhouse of the earth, allowing sunlight to reach the earth's surface but limiting reradiation of the resulting heat into space. The temperature of the earth—which profoundly affects the suitability of the environment for life—is therefore certain to rise as the amount of carbon dioxide in the air increases.

A report by the President's Science Advisory Committee[1] finds that the extra heat due to fuel-produced carbon dioxide accumulated in the air by the year 2000 might be sufficient to melt the Antarctic ice cap—in 4000 years according to one computation, or in 400 years according to another. And the report states: "The melting of the Antarctic ice cap would raise sea level by 400 feet. If 1,000 years were required to melt the ice cap, the sea level would rise about 4 feet every 10 years, 40 feet per century." This would result in catastrophe for much of the world's inhabited land and many of its major cities.

A more recent energy source—the internal-combustion engine—is polluting the environment much faster than fire. The automobile is only about seventy years old, but in that time it has severely damaged the quality of the air. The air over most large cities carries a large burden of waste automobile fuel. On exposure to sunlight this forms the noxious ingredients of smog, which significantly increases the incidence of respiratory disease. Since tetraethyl lead was introduced in 1923 as an automobile fuel additive, lead has contaminated most of the earth's surface. Increasing amounts of the metal are found in surface ocean waters, in crops, and in human blood, in which in some areas the amount may be approaching toxic levels.

As large a body of water as Lake Erie has already been overwhelmed by pollutants and has, in effect, died. In its natural state, Lake Erie was a balanced system in which water plants, microorganisms, and a great variety of swimming creatures lived together in an intricate harmony. But today most of Lake Erie is dead. Sewage, industrial wastes, and the runoff from heavily fertilized farmlands have loaded the waters of the lake with so much excess phosphate and nitrate as to jar the biology of the lake permanently out of balance. The fish are all but gone. According to a recent report by a committee of the National Academy of Sciences, within about twenty years city wastes are expected to overwhelm the biology of most of the nation's waterways.

Small amounts of nitrate are naturally present in all bodies of water, and living things can tolerate—and often require—nitrate at these low levels. Now, however, nitrate originating in the outflow waters of sewage-treatment plants and in the runoff from land treated with chemical fertilizers has begun to build up excessively in ground water in thirty-eight regions of the United States, according to a recent Geological Survey report. Excess nitrate is poisonous to man and animals. About 8 to 9 parts per million of nitrate in drinking water causes a serious respiratory difficulty in infants—cyanosis—by interfering with hemoglobin function. For domestic animals 5 parts per million is considered unsafe. Some wells in the United States already have more than 3 parts per million of nitrate, and the contamination levels will go up with increased use of fertilizers and the growing density of population.

Added to the growing volume of the more

familiar wastes are numerous new pollutants, produced through the ingenuity of modern physics and chemistry: radioactive elements, detergents, pesticides, weed killers, and a variety of industrial wastes. The greatest single source of contamination of the planet is now the radioactivity from test explosions of nuclear weapons in the atmosphere.[2] Fallout from nuclear tests contaminates every part of the earth's surface and all of its inhabitants. Strontium-90, one of the radioactive constituents of fallout, is being built into the bones of every living person and will be carried in the bodies of several future generations. The fallout problem can tell us a good deal about the connection between modern science and the hazards of life on the earth.

Contamination of the earth's surface with fallout originates in the scientific revolution set off fifty years ago by far-reaching discoveries in atomic physics. By 1940 it was apparent that the new knowledge of atomic structure could lead to technological processes of vast power and scope. That these potentialities were so rapidly realized reflects the force of military demands. Faced with the grim dangers of war with Nazi Germany, the governments of the United States and Great Britain undertook the monumental task of translating what was until then an esoteric laboratory experiment—nuclear fission—into the awesome reality of the nuclear bomb. The bomb was created by the magnificent new insights of nuclear physics, driven to success by the determination to apply the full force of modern science to victory in the war.

With later scientific advances, nuclear weapons of increasing explosive power became possible.

Propelled by the fears and tension of the cold war, these possibilities were fully exploited by those nations capable of making the necessary economic and technological effort—the United States, the U.S.S.R., Great Britain, and to a lesser extent, France and China. As a result, in 1948 there began a constantly accelerating series of nuclear explosions designed to develop weapons of increasing destructiveness and versatility. The total explosive power released by nuclear-weapon tests between 1948 and 1962 was equivalent to about 500 million tons of TNT—nearly two hundred times the total power of all the bombs dropped on Germany in World War II. The amount of only one fallout component—strontium-90—released by nuclear tests has introduced into the environment radioactivity equivalent to about 1 billion grams of radium. The significance of this sudden radioactive intrusion can be visualized by comparing it with the world supply of radium before World War II—about 10 grams. Until the advent of nuclear fission these few grams of radium represented the total human experience in handling radioactive substances.

The rapid and unchecked expansion of nuclear weaponry, which now dominates the world's major military programs, testifies to the enormous success of nuclear physics and engineering. There have been no complaints about the power and reliability of nuclear weapons. But official appraisals of the *biological* consequences of nuclear explosions have undergone a drastic change. Contrast statements, only eight years apart, by two American presidents:

President Eisenhower, October 1956: "The continuance of the present rate of H-bomb testing, by

the most sober and responsible scientific judgment
... does not imperil the health of humanity."[3]

President Johnson, October 1964: "This treaty
[the nuclear test-ban treaty] has halted the steady,
menacing increase of radioactive fallout. The
deadly products of atomic explosions were poisoning our soil and our food and the milk our children drank and the air we all breathe. Radioactive deposits were being formed in increasing
quantity in the teeth and bones of young Ameri-
cans. Radioactive poisons were beginning to threat-
en the safety of people throughout the world. They
were a growing menace to the health of every un-
born child."[4]

There were, of course, some political reasons
for this dramatic reversal in government policy—
even President Eisenhower softened his rejection
of the nuclear test ban before he left office. But it
is now quite clear from available documents that
between 1956 and 1963, when the U.S.—Soviet
nuclear-test-ban treaty was signed, there were also
sharp revisions in our appraisal of the fallout hazard.

Evaluation of this problem calls for a detailed
analysis of the passage of radioactive isotopes—for
example, strontium-90—from their creation in the
nuclear fireball to their entry into the human
body. When formed, strontium-90 and other radioactive elements become attached to tiny dustlike
particles. These are shot high into the stratosphere by the explosion and return to earth at a
rate which depends on the size of the particle
and on the weather. Eventually the fallout is carried to the earth in rain or snow. From that point
its fate is determined by the complex biology of
soil, plants, domestic animals, and finally man. For

example, since strontium-90 is a chemical relative of calcium, it takes a similar biological course. Along with calcium and other minerals, strontium-90 enters grass and crop plants through their roots and from raindrops resting on their leaves. When cows eat the contaminated grass the strontium-90 concentrates in their milk, which is rich in calcium. People absorb strontium-90 from contaminated milk and other foods. Once the amount of fallout absorbed in the body is known, its radiation output can be calculated and the resultant medical risk of radiation-induced cancer or genetic change can be estimated.

Our Faulty Knowledge of Fallout

To understand the biological effects of fallout we must know what happens at each step of this complicated chain of events. There is now a fairly complete published record of this knowledge. This record reveals a number of important errors in our understanding of the problem, which remained uncorrected until the testing of nuclear weapons in the atmosphere was in full swing and fallout had been massively disseminated into the environment.[5]

In 1953, AEC publications asserted that fallout would be evenly distributed over the globe so that no area would receive an excessive amount. In 1958 actual measurements of fallout distribution showed that this idea was wrong. Fallout levels in the North Temperate Zone were found to be more than ten times higher than at the equator or the poles and five times higher than in the South Temperate Zone. Because most of the world's population lives in the North Temperate Zone, the total human exposure to fallout is much greater than the original AEC theory anticipated.[6]

As late as 1957, the official government handbook, *The Effects of Nuclear Weapons,* published jointly by the AEC and the Department of Defense, claimed that fallout would descend from the stratosphere slowly, half of it not reaching the earth until seven years after it was produced. This delay, it was believed, would allow time for the harmless decay of a large proportion of the radioisotopes in fallout while they were still at high altitudes, thus minimizing the ultimate exposure of humans. In 1962 the second edition of this same handbook acknowledged that this estimate was wrong. Most of the stratospheric fallout descends to the earth in a matter of months, depending on the geographic location and the time of the explosion. A considerable amount of hazardous radioactivity remains in fallout when it reaches the earth.

As a result of these errors, early estimates of the amount of fallout that would reach the soil of the United States were far off the mark. In 1956 AEC Commissioner W. F. Libby predicted that nuclear tests carried out through May 1954 would deposit in United States soil a maximum of 7 millicuries of strontium-90 per square mile. But actual measurements by the AEC showed that the average strontium-90 content of U.S. soil had reached about 47 millicuries per square mile in 1958, although the total amount of testing had only about doubled since 1954.[7]

Mistakes were also made in evaluating the path that fallout radioisotopes would take through the food chain to man. In 1953 the AEC stated that the only possible hazard to humans from strontium-90 would arise from "the ingestion of bone splinters which might be intermingled with muscle tissue during butchering and cutting of the meat."[8] No mention was made of the simple biological fact that milk from such an animal would also contain strontium-90. By 1956 the AEC had acknowledged that milk represented the most important source of strontium-90 in human food.

Important mistakes were also made in judging the medical hazards of fallout. For example, the risk of genetic damage from fallout was at first dismissed by a 1953 AEC report with a statement that: "Fallout radioactivity is far below the level which could cause a detectable increase in mutations, or inheritable variations."[9]

By 1957 a report of the AEC Biological and Medical Advisory Committee had concluded that fallout from tests completed to that date would probably result in 2,500 to 13,000 cases of serious genetic defects per year throughout the world population.

One of the major mistakes made in evaluating the fallout hazard was to rely on the *average* values of fallout exposure, for such an average can conceal areas in which special circumstances combine to intensify the danger. For example, in the arctic only slight radiation exposure was expected, because the amount of fallout that reaches the ground at the poles is much less than it is in the temperate parts of the United States. However, Eskimos in the arctic have now been discovered to have amounts of fallout radioactivity in their bodies which are much higher than those found in the inhabitants of temperate regions. The clue to this puzzle was finally found in the special nutrition of lichens—a group of composite organisms, each of which consists of an alga and a fungus living together. Lichens, which have no functional roots and often grow on rocks rather than soil, absorb their mineral nutrition in the form of dust taken directly from the air. Thus natural "fallout" is the lichen's chief source of minerals. Radioactive fallout also reaches lichens from the air, and these plants are not protected by the dilution and discrimination processes which operate when fallout is absorbed through roots. Caribou eat great quantites of lichen and therefore take up excessive amounts of fallout. At the end of the food chain, caribou meat makes up a considerable part of the Eskimos' diet, producing the unexpectedly large amounts of fallout radiation in their bodies. The operation of this special food chain in the arctic upset the expectations based on temperate-zone food chains about the amount of fallout which can find its way into the diet.

In one sense the nuclear test program must be regarded as a remarkable scientific triumph, for it solved very difficult physics and engineering problems. But the biological consequences of the nuclear test program—the vast intrusion of radioactivity into animals, plants, and man—must be regarded as a huge technological mistake. There have been serious oversights and miscalculations. It is now clear that the government agencies responsible for the development of nuclear weapons embarked on this massive program before they understood the full biological effects of what they proposed to do. Great amounts of fallout were disseminated throughout the world before it became known that the resultant medical risks were so great as to require that nuclear testing be halted. The enactment of the test-ban treaty in

1963 is, in part, a confession of this failure of modern science and technology.

Detergents

A different kind of pact—an agreement among the companies that make up the multibillion-dollar United States detergent industry—was needed to correct another big technological mistake. These companies agreed to replace, by July 1, 1965, the main active ingredient which, beginning in the early 1940's, had enabled their products to capture the major share of the cleanser market.

Detergents are chemicals, synthesized from raw materials found in petroleum, which have largely replaced soap in many household and industrial uses. Soap is itself one of the earliest-known useful chemicals. Long ago, it was discovered that fats and oils extracted from animals or pressed from seeds, and cooked with alkali, react chemically to form soap. Soap is a double-headed molecule. Its fatty part readily combines with droplets of greasy dirt, forming a film around them. The other end of the molecule forms the outer layer of the soap-coated dirt particle and has a strong affinity for water. As a result, the whole complex can be washed away.

But soap has certain technical and economic disadvantages. In "hard" water, which has a high mineral content, soap forms a desposit which will not readily wash away. In addition, the raw material, fat, is dependent on agriculture, hence variable in quality, availability, and price.

In the 1930s chemical technology began to produce synthetic organic compounds which closely resemble natural products: synthetic fibers, plastics, and artificial rubber. The inadequacies of soap were an attractive challenge for chemical engineers, who set out to make a synthetic washing agent that might avoid the hard-water problem and be independent of the vagaries of agriculture.

Detergent research began with fatlike molecules —hydrocarbons—which are common in petroleum. Chemists found ways to attach to the hydrocarbon a water-soluble molecular group containing sulfur. The result was a family of substances which, like soap, formed a water-soluble coat around grease particles. But the new detergents were better than soap, since they were equally effective in hard or soft water. Intensive research produced detergents with other useful properties: long shelf life; pleasantness to touch, sight, and smell; gentleness on the hands; economical price. This success illustrates the effectiveness of modern chemical research.

Within a few years after the new detergents were placed on sale, they had won a very large portion of the cleanser market. By 1960 they had replaced soap as the major household and industrial cleanser. A new billion-dollar industry had been created. However, one aspect of this technological triumph received no attention in the research laboratories—the effects of dumping a huge amount of new synthetic substances (about 3.5 billion pounds per year in the United States in 1960) down drains into waste-disposal systems. This lack of interest was perhaps natural, since the purchases of detergents—and the consequent profits—result from their effectiveness as cleansers and not from their behavior in waste systems. Finally it became difficult to ignore this aspect. Mounds of detergent foam appeared in riverways; in some places a glass of water drawn from a tap foamed up a head that would make a brewer envious. Only then was it discovered that, despite their useful similarities to soap, the new detergents differed from the natural product in one important way. When soap enters waste-disposal systems, it is readily broken down by bacteria, but the detergents are not. They pass through the waste system unchanged, appearing in the runoff water that drains into rivers, streams, and underground waters. Water supplies taken from these sources therefore contain detergents.

Now, for the first time, industrial chemists were forced to investigate another aspect of detergent chemistry: Why do detergents resist natural degradation? The difficulty was found to arise from the structure of the hydrocarbon skeleton of their molecules. Hydrocarbon molecules consist of a chain of carbon atoms, to each of which are attached two or three hydrogen atoms. The troublesome detergents had branched chains, and it was discovered that the bacterial enzymes which readily break up natural unbranched hydrocarbons in sewage plants are unable to degrade the synthetic branched hydrocarbons.

By the 1960's, water contamination had become bad enough to stimulate legislative action—and, simultaneously, the needed research. Resulting knowledge of the basic cause of the difficulty suggested a possible remedy. Methods were developed for producing degradable detergents from the unbranched hydrocarbon molecules that are also found in petroleum. The industry agreed to replace the branched-molecule detergents with branchless ones by July 1, 1965.

This solves part of the problem. In the large

well-aerated sewage systems of most urban communities the new detergents are destroyed by bacteria. However, they do not break down as readily in underground unaerated systems, such as septic tanks. About 34 percent of the homes in the United States are equipped with septic systems. Since half of these also obtain water from their own wells, even the new "degradable" detergents may cause trouble. Moreover, when they are degraded the new synthetic detergents may overload surface waters with phosphate, leading to serious upsets in biological balance, such as the disaster which has already overtaken Lake Erie.

Thus, long after synthetic detergents had become a common household item, they were found to cause an intolerable nuisance in water supplies. There is no way to gloss over this episode. It represents a failure on the part of modern chemical technology to predict a vital consequence of a massive intervention into nature.

Insecticides

My final example is one from personal experience. During World War II, I served as project officer in the Navy's development of aircraft dispersal of DDT, which proved to be of great importance in the Pacific island battles by protecting the first wave of attackers from serious insect-borne diseases. The project required meticulous studies of aerosol production, aerodynamic distribution of insecticide droplets, insect kill, meteorological effects, and the problems of flying tactics. Toward the end of our work, when the system was ready for fleet operations, we received a request for help from an experimental rocket station on a strip of

island beach off the New Jersey coast. Flies were
so numerous on the beach that important military
developments were being held up. We sprayed the
island and, inevitably, some of the surrounding
waters with DDT. Within a few hours the flies
were dead, and the rocketeers went about their
work with renewed vigor. But a week later they
were on the telephone again. A mysterious epidemic had littered the beach with tons of decaying
fish—which had attracted vast swarms of flies from
the mainland. This is how we learned that DDT
kills fish.

Such unexpected twists are often encountered
when new synthetic substances are thrust into the
complex community of life: a wholly unanticipated development wipes out their original usefulness, or sometimes creates a problem worse
than the original one. In one Bolivian town, DDT
sprayed to control malarial mosquitoes also killed
most of the local cats. With the cats gone, the
village was invaded by a wild, mouselike animal
that carried black typhus. Before new cats were
brought to restore the balance, several hundred
villagers were killed by the disease.[10]

The Scientific Background
of Technological Failures

These problems have a common scientific background. Each of them springs from a useful technological innovation. The burning of fuel by
internal-combustion engines is an enormously valuable source of energy—but also pollutes the air.
New synthetic chemicals, the fruits of remarkable
advances in chemical technology since World War
II, appear in a multitude of useful forms—but also

as new pollutants of air and water. The development, about twenty-five years ago, of self-sustained nuclear reactions has given us not only new weapons and new sources of power, but unprecedented radioactive debris as well.

Most of these problems seem to crop up unexpectedly. The sunlight-induced chemical conversion of airborne hydrocarbons (such as gasoline vapor) into smog was discovered, not in a chemical laboratory but in the air over Los Angeles, long after the chief mode of disseminating these hydrocarbons—the superhighway—was well entrenched in the urban economy. The full significance of the absorption of fallout into the human body became known only some years after the establishment of massive programs of nuclear testing. Most of the medical hazards of the new insecticides were noticed only long after these substances were in wide use. All these problems have been imposed on us—sometimes to our considerable surprise—well after the causative activity was in full swing.

Could we cure these difficulites by calling a halt to science and new technologies? The present accelerating growth of science and technology—which, together with population growth, is the cause of most of our pollution problems—was set in motion more than sixty years ago. Its roots are in the scientific revolution which took place at the turn of the century, when physicists discovered that the apparently simple laws of Newton's time concealed a complex world of exceedingly small particles and immense forces. From this knowledge has come the great flowering of modern science—including the new energy sources and syn-

thetic substances which have covered the earth
with pollution. We are today witnessing the inevitable impact of the tidal wave created by a scientific revolution more than half a century old. It is
simply too late to declare a moratorium on the
progress of science.

The real question is not *whether* we should
use our new knowledge, but *how* to use it. And to
answer that, we must understand the structure of
modern scientific knowledge: in which areas the
new insights of science are powerful and effective
guides to action; in which others they are too
uncertain to support a sound technology. Since the
scientific revolution which generated modern
technology took place in physics, it is natural that
modern science should provide better technological control over inanimate matter than over living
things. This disparity is evident in our environmental problems. If basic theories of physics had
not attained their present ability to explain nuclear structure, we would not now be confronted
with massive dissemination of manmade radioisotopes and synthetic chemicals. If biological theory
had become sufficiently advanced to master the
problems of cancer—a chief hazard from modern
pollutants—we might be better prepared to cope
with these new environmental contaminants. We
are in difficulty because of the wide disparity between the present state of the physical and the
biological sciences.

The separation of the laws of nature among the
different sciences is a human conceit; nature itself
is an integrated whole. A nuclear test explosion is
usually regarded as an experiment in engineering
and physics; but it is also a vast, if poorly con-

trolled, experiment in environmental biology. It is a convincing statement of the competence of modern physics and engineering, but also a demonstration of our poor understanding of the biology of fallout. If the physicochemical sciences are to be safely used in the new technologies they will need to be governed by what we know—and do not know—about life and its environment.

The Biosphere

What we know about living things and about the biosphere—the community of life in the environment—is that they are enormously complex, and that this complexity is the source of their remarkable staying power. The web of relationships that ties animal to plant, prey to predator, parasite to host, and all to the air, water, and soil which they inhabit persists *because* it is complex. An old farmhouse practice is a simple illustration of this fundamental point. Farmers who keep cats to control the ravages of mice find it necessary to offer the cats a doorstep feeding. Only if the farmer provides this alternative source of food can the cats withstand a temporary shortage in the mouse supply and remain on hand to catch the mice when they reappear. A stable system that will keep mice in check must comprise all three components: cats, mice, and domestic cat food. This principle is well established in environmental biology: anything which reduces the complexity of a natural biological system renders it less stable and more subject to fatal fluctuations.

The biosphere is closely governed by the connections among its numerous parts. The connections which comprise the biological food chain, for example, greatly amplify the effects of environmental pollution. If soil contains 1 unit of insecticide per gram, earthworms living in the soil will contain 10 to 40 units per gram, and in woodcocks feeding on the earthworms the insecticide level will rise to about 200 units per gram. In the biosphere the whole is always greater than the sum of its parts; animals which absorb one insecticide may become more sensitive to the damaging effects of a second one. Because of such amplifications, a small intrusion in one place in the environment may trigger a huge response elsewhere in the system. Often an amplification feeds on itself until the entire living system is engulfed by catastrophe. If the vegetation that protects the soil from erosion is killed, the soil will wash away, plants will then find no footholds for their seeds, and a permanent desert will result.

It is not surprising, then, that the introduction of any killing chemical into the environment is bound to cause a change somewhere in the tangled web of relationships. For this reason, and because we depend on so many detailed and subtle aspects of the environment, *any* change imposed on it for the sake of some economic benefit has a price. For the benefits of powerful insecticides we pay in losses of birdlife and fish. For the conveniences of automobiles we pay in the rise of respiratory disease from smog. For the widespread use of combustible fuels we may yet be forced to pay the catastrophic cost of protecting our cities from worldwide floods. Sooner or later, wittingly or unwittingly, we must pay for every intrusion on the natural environment.

Our Knowledge Is Dangerously Incomplete

There is considerable scientific disagreement about the medical hazards of the new pollutants: about the effects of DDT now found in human bodies, about the diseases due to smog, or about the long-range effects of fallout. But the crucial point is that the disagreements exist, for they reveal that we have risked these hazards before we knew what harm they might do. Unwittingly we have loaded the air with chemicals that damage the lungs, and the water with substances that interfere with the functioning of the blood. Because we wanted to build nuclear bombs and kill mosquitoes, we have burdened our bodies with strontium-90 and DDT, with consequences that no one can now predict. We have been massively intervening in the environment without being aware of many of the harmful consequences of our acts until they have been performed and the effects—which are difficult to understand and sometimes irreversible—are upon us. Like the sorcerer's apprentice, we are acting upon dangerously incomplete knowledge. We are, in effect, conducting a huge experiment on *ourselves*. A generation hence—too late to help—public health statistics may reveal what hazards are associated with these pollutants.

To those of us who are concerned with the growing risk of unintended damage to the environment, some would reply that it is the grand purpose of science to move into unknown territory, to explore, and to discover. They would remind us that similar hazards have been risked before, and that science and technology cannot make progress without taking some risks. But the size and persistence of possible errors has also grown with the power of science and the expansion of technology. In the past, the risks taken in the name of technological progress—boiler explosions on the first steamboats, or the early injuries from radium—were restricted to a small place and a short time. The new hazards are neither local nor brief. Air pollution covers vast areas. Fallout is worldwide. Synthetic chemicals may remain in the soil for years. Radioactive pollutants now on the earth's surface will be found there for generations, and, in the case of carbon-14, for thousands of years. Excess carbon dioxide from fuel combustion eventually might cause floods that could cover much of the earth's present land surface for centuries. At the same time the permissible margin for error has become very much reduced. In the development of steam engines a certain number of boiler explosions were tolerated as the art was improved. If a single comparable disaster were to occur in a nuclear power plant or in a reactor-driven ship near a large city, thousands of people might die and a whole region be rendered uninhabitable—a price that the public might be unwilling to pay for nuclear power. The risk is one that private insurance companies have refused to underwrite. Modern science and technology are simply too powerful to permit a trial-and-error approach.

It can be argued that the hazards of modern pollutants are small compared to the dangers associated with other human enterprises. For us, today, the fallout hazard is, for example, much smaller than the risks we take on the highway or in the air. But what of the risks we inflict on future generations? No estimate of the actual harm that

3

Greater Than the Sum of Its Parts

As we have seen, many of our recent technological mistakes crop up as an unexpected biological aftermath of a new advance in physics or chemistry. Our present grasp of biology appears to be inadequate to explain what happens when living things encounter radiation or the new synthetic chemicals. But this concept conflicts with reports of recent sweeping advances in our basic understanding of life. The claim has been made, for example, that twentieth-century science will be remembered more for its achievements in biology than for nuclear physics.

Such claims reflect a conviction that the basic laws of biology have now been disclosed and, in particular, that we now know that life is a form of chemistry. In this view the undoubted ability of the modern physicochemical sciences to understand the workings of inanimate matter ought to give them the same power to penetrate the mysteries of life. The main basis for claims that we now know "the secret of life" are recent discoveries about the chemical processes involved in the unique features of life—growth, reproduction, and inheritance. It is worth examining the background of these new ideas for some explanation of the paradoxical discrepancy between our technological failures and our apparent understanding of the basic laws of life.

Two Kinds of Biology

Anyone who learned biology by dissecting a frog must find the reports of present-day biological research strange and unsettling: molecules that reproduce themselves; a molecular "code" that tells an egg whether it should turn into a turtle or a tiger; efforts to create life in a test tube of chemicals.

These new ideas seem to clash with long-familiar principles of biology. If a molecule possesses the essential property of life—self-duplication—then the cell theory, which states that the attributes of life reside in the whole cell and not in any smaller part, such as a molecule, must be abandoned. If test-tube synthesis of life were to be achieved, the hitherto unchallenged principle, *omnes ex ovo*—that all life comes from pre-existing life—would have to be given up. If all the features of the adult are encoded in the fertilized egg, we must revive the supposedly disproved idea of "preformation," the molecular code taking the place of the tiny fetus which early microscopists imagined they saw curled up in the human egg or sperm.

Are the older principles of biology now outmoded by the new research? If so, modern science has made a major revolution in biological theory. On the other hand, if this interpretation of the new advances is wrong, that might help explain why we have blundered into biological difficulties in our recent technology.

These questions reflect two conflicting concepts about the nature of life: classical biology and "molecular" biology. Classical biology is built upon observations and experiments with authentic living organisms, organs, and cells. Classical biology insists upon studying these complex systems because no simpler ones are alive; it assumes that life is inherently associated with the complexity of at least the whole cell. The approach which we now call molecular biology, however, assumes that chemical constituents, separated from the cell and studied with sufficient subtlety and detail, will be found to possess properties which, in themselves, can account for life. It suggests that life might even reside in a single cellular constituent and permits the notion of a "living molecule."[1]

We are sometimes told that molecular biology is a modern science, while classical biology is a surviving relic of the nineteenth century. This view casts the conflict into a familiar pattern: old-fashioned classical biology doggedly resisting inevitable replacement by the up-to-date molecular variety. Despite its convenient simplicity, this view is inaccurate, for both classical and molecular biology have long and intimately connected histories. They represent two paths which started nearly two hundred years ago and have steadily converged toward their present collision.

The Classical Approach

Classical biology began with efforts to describe and classify living things as they are found in nature—in full possession of their diverse faculties, and alive. Biological classification, which arose from the practical necessity of cataloguing and naming newly discovered plants and animals, was placed on a scientific basis by Linnaeus in the eighteenth century. The success of Linnaeus's system of relationships among the numerous animal and plant species becomes clear when compared with one of his failures. Linnaeus also established a system of geological classification based on the obvious properties of rocks, but this system failed as soon as the advance of chemistry described the molecular composition of rocks. In contrast, the Linnaean system of biological classification, although also based on readily discernible features of the whole organism, has, with relatively minor changes, withstood the now vastly detailed information about the chemical composition of different living things. A botanist with only a hand lens can often discern the distinguishing characteristics of a plant species as well as or better than a biochemist armed with an entire analytical laboratory. This achievement belies the notion that classification and naming of species is a dull and superficial aspect of biology more suited to the minds of children and retired country gentlemen than to the subtle intellect of the scientist.

Other achievements of classical biology which, like the Linnaean system of classification, were also based on the study of intact living organisms are the principles of evolution and the Mendelian laws of genetics. By the turn of the century classical biology had developed, from observations of living organisms, a system of knowledge which characterized the diversity of life, delineated the origins of species in the natural selection of inborn variations, and described how variation is transmitted from one organism to anotther.

With the development of the microscope, the gross features of plants and animals were revealed as the natural consequence of an internal structure based on a universal unit, the cell. The cell appeared to be the seat of life. Separated from the organism the cell still retains the features of vitality: metabolic transformations of food substances, responsiveness to environment, energetic activity, reproduction, and inheritance. But when the cell is dismembered, these capabilities are lost, even though certain isolated parts, such as enzymes, are able for a time to carry out one or another chemical reaction.

The microscope also led to important discovereies about the development of living things. When the first microscopes were built, it could be seen that even an animal as marvelously contrived as man begins life as a single cell. Searching for the source of the detailed structures which later appear in the adult, some early microscopists fancied they saw a tiny fetus, or homunculus, in the egg or sperm. But the homunculus, and with it the theory of preformation, disappeared when the power of the microscope was improved. It became clear that, although an animal's development begins with the relatively simple spherical fertilized egg cell, the new cells arising from the egg's division become differentiated, eventually taking on their special appearance and functions: skin cells, broad and flat; muscle cells, long and contractile; nerve cells with extended conductile fibers. The science of biological development, embryology, describes this course of events. It shows that plants and animals are produced by a closely coordinated series of events, which steadily adds to the organism's structural detail. This leads in turn to the concept of *epigenesis*—that is, that development is due to the emergence of new structures rather

than to the unfolding of preformed miniature
parts.

The Molecular Approach

The second path of biological research is nearly as
old as classical biology. It is concerned with the
chemistry of the cell—what molecular constituents
the cell contains and how they react with one
another. These investigations may start with the
living organism, or cell, but only in the sense that
the recipe for rabbit stew begins with a rabbit.
Most constituents cannot be studied until they are
removed from the cell, an operation which, like
cookery, renders the cell—or the rabbit—dead.

This new chemistry has been of enormous importance for the understanding of living things. It
has defined the chemical nature of foodstuffs, determined what substances are essential for the construction of cellular components, and described
the chemistry of the energetic processes which
power the cell. Gradually, piece after piece of the
cell's structure and chemical machinery have been
isolated, purified, and meticulously studied as to
size, shape, composition, and chemical reactivity.
These efforts, especially since the turn of the century, have created a new science, biochemistry,
which has developed *molecular* explanations for
many of the special capabilities of living organisms.

For example, early biochemical research showed
that the oxygen-carrying power of the blood is due
to hemoglobin, which consists of an iron-
containing red pigment associated with a large
protein constituent. Later, studying how X-ray
beams were scattered from artificial crystals of he-

moglobin, crystallographers could describe the
molecule's intimate structure and show how the
pigment is tucked away in the convoluted folds of
the protein. The special capability of the blood to
provide tissues with oxygen could then be related
to the particular atomic arrangement of the iron
in the pigment, and of the pigment in the protein.
This aspect of life was given a molecular explanation.

No conflict between the two kinds of biology is
evident here. While molecular biology localized
the unique function of the blood in the intricate
structure of hemoglobin, the classical biologist
could reply that this marvelous molecular machine itself is created only by the living cell.
Hence the principle of classical biology which
holds that the special capabilities of life belong to
nothing less complex than the cell remained intact.

Inevitably, biochemistry took the next step and
succeeded in isolating from cells the enzymes and
other cellular constituents which appear to cat-
alyze the synthesis of proteins, such as hemoglobin. Even then, it could be argued that although the synthetic machinery has been isolated
from the cell, the *guidance* of this machinery—
which determines what kind of protein is to be
synthesized—remained the property of the whole
cell.

Then, inexorably, the classical biologist was
denied even this last refuge. In the last few years
it has been shown that a separable constituent of
the cell—deoxyribonucleic acid, or DNA—appears
to guide the cell's protein-synthesizing machinery.
The internal structure of DNA seems to represent
a set of coded instructions, which, through an in-

termediary (ribonucleic acid or RNA) , dictates the pattern of protein synthesis, much as a punched tape can determine whether a modern machine tool will shape a metal blank into a cam or a gear. And other experiments seem to show that in the presence of appropriate enzymes each DNA molecule can form a replica, a new DNA molecule, containing the specific guiding message present in the original. This idea, when added to what was already known about the cellular mechanisms of heredity (especially that DNA is localized in the chromosomes, which carry many inherited factors) , appeared to establish a molecular basis for inheritance itself. Such results serve to justify the new term now applied to physicochemical analysis of life: molecular biology.[2]

Is DNA the Secret of Life?

Molecular biology has indeed separated from the cell a series of components which, of themselves, seem to accomplish the unique chemical events associated with life: metabolic catalysis; synthesis of simple molecules, and complex ones such as proteins and nucleic acids; guidance of synthesis; and finally, inheritance. If replication and inheritance are the source of all the other features of life, then DNA emerges as "the master chemical of the cell," and "the secret of life."

I have stated these results in the form of broad conclusions—the generalizations one reads in news magazine reports, often in textbooks, and sometimes in scientific papers. They are satisfyingly simple and understandable and would seem to mark a clear end to the long-debated theoretical question of whether the uniqueness of life originates in the whole cell or in some molecular constituent.

But the apparent victory of DNA is not the result of a successful resolution of the long debate. Instead the decision has been reached by the less arduous expedient of largely ignoring the basic theoretical question. This question has not really been answered by the new experiments, for they do not in fact support the idea that DNA is a "self-duplicating molecule."

Crucial experimental evidence for the conclusion that DNA is a self-duplicating molecule would require a demonstration, in a test-tube system supplied with a particular DNA molecule, of the formation of new DNA molecules which are the exact replicas of the original one. But this has not yet been accomplished. Although such test-tube systems do synthesize new DNA that *resembles* the DNA used to prime the process, the resemblance is by no means as exact as the theory requires. Recent experiments show that a test-tube system primed with a simple two-ended fiber of DNA produces new DNA strands much longer than the original, and highly branched. More significantly, the enzyme which synthesizes DNA can sometimes produce a specific DNA in the absence of a starting DNA molecule. This suggests that the hereditary specificity of DNA is derived not only from a parental DNA molecule but also from the enzyme that synthesizes it. In this case, DNA cannot properly be regarded as a self-duplicating molecule.

Recent reports of experiments on the test-tube replication of a biologically active nucleic acid from a virus which infects bacteria show the same interdependence of nucleic acid and the enzyme which synthesizes it. In these experiments, virus nucleic acid appears to be replicated quite exactly in the test tube—but only if this process is guided by a very specific enzyme extracted from an already virus-infected bacterium.

Test-tube experiments on the guided synthesis of proteins reveal a similar multiplicity in the replicative process. These experiments do show that the chemical composition of the nucleic acid has an influence on the chemical composition of the protein that is made. This is the basis of the so-called code by which the genetic information in nucleic acid is presumed to govern the chemistry of protein synthesis, and thus determine the inherited characteristics of the organism. However, recent test-tube experiments show that other agents beside the nucleic acid also have a guiding influence. The kind of protein made may depend on the specific organism from which the necessary enzyme is obtained. It also depends on the test tube's temperature, the degree of acidity, and the amount of metallic salts present. If the nucleic-acid code spells out instructions for inheritance, then, according to these results, it gives different information at different temperatures. Of course, in the actual organism inheritance is not so flexible. What the test-tube experiments tell us is that there is no single molecular message which determines the inheritance of the living organism. The beautiful exactness of biological inheritance depends on the precise interactions of many molecular processes. The message which controls heredity is not carried by a single molecule but by the whole living cell.

What is at issue here is not the validity of the experimental results which describe the structure and properties of DNA, RNA, and proteins, but the sweeping generalizations which these results have engendered. These generalizations are unsound, I believe, not because they are founded on faulty data, but because they do not take into account *all* the relevant data and are based on an arbitrary exclusion of certain essential facts. Certainly there is good experimental evidence to support the idea that DNA does influence the hereditary characteristics of living cells in which it occurs, but this is also true of other cellular constituents, such as the enzyme that synthesizes DNA. The basic question is whether DNA represents a self-contained code that *by itself* determines whether the organism is a turtle or a tiger. Such a conclusion is based not on experimental fact, but on dogma.

If you are shocked by my use of this word, let me hasten to add that it is not my own. The term "central dogma" is often used in current scientific literature to describe the principles which are supposed to explain the governing role of DNA in inheritance. At the annual meeting of the Federation of American Societies of Biology in April 1965, for example, a symposium on the inheritance of disease was opened by the chairman with the statement: "Let me review quickly the essential doctrines which comprise the dogma of modern genetics."[3]

In the index to a report of another symposium on molecular genetics the entry "Dogma, The" is followed by eleven page references. At one point a symposium participant makes this remarkable statement: "The reason we call this 'dogma' is that it depends on personal bias, not logic."[4]

The terms "dogma" and "doctrine" were introduced into the scientific literature by proponents of the DNA code theory. When the "central dogma" was first enunciated by one of the founders of the DNA code theory, his choice of the title was probably seen as a kind of in-group joke. But the humor concealed a serious intellectual act. The "central dogma" involved an admittedly unproven assumption: that while nucleic acids can guide the synthesis of other nucleic acids and of proteins, the reverse effect is impossible—that is, proteins cannot guide the synthesis of nucleic acids. But actual experimental observations deny the second, and crucial, part of this assumption. Certainly nucleic acids can contribute to the guidance of the synthesis of proteins, that is, help determine the exact order in which its constituent parts—amino acids—appear in the protein fiber. But at the same time, some proteins—for example, the enzyme that synthesizes DNA—can contribute to the guidance of the synthesis of nucleic acid and help to determine the order in which *its* constitutes—nucleotides—appear in the nucleic-acid fiber. Neither the nucleic acid nor the protein *alone* is capable of such guidance of synthetic chemistry. No one molecule is "self-duplicating," or the "master chemical" of the cell.

By means of an arbitrary fiat the "central dogma" banishes from consideration the interactions among the numerous molecular processes that have been discovered in cells or in their extracted juices. And like other dogmatic decrees this one too flies in the face of reality. In the real living cell, molecular processes—synthesis of nucleic acids and proteins or the oxidation of food substances—are not separate but interact in exceedingly complex ways. The simple sum of the separate molecular events is insufficient to represent the living whole. Some subtle cellular property is lost when, in the first step of a biochemical procedure, the cell is broken open and killed. The best evidence for this assertion is a rather simple fact: No matter how many ingredients they may contain, the biochemist's test-tube mixtures are dead; but these same ingredients, organized by the subtle structure of the cell, comprise a system which is alive.

The Whole and Its Parts

A common reaction to this kind of suggestion is that it is an unhealthy resurgence of vitalism, which proposes that some nonmaterial "vital force" wholly unsusceptible to scientific analysis adds "life" to the otherwise lifeless substance of the cell. But a single example from another field of science shows that complex systems may, for perfectly natural reasons, exhibit distinctive properties which are not at all discernible in the behavior of their isolated parts.

Consider the recent successes in the analysis of the unique electrical property exhibited by metals at very low temperatures: superconductivity. At ordinary temperatures the flow of electricity in a metal is governed by the familiar Ohm's law, which relates current flow, resistance to flow, and the strength of the driving electromotive force. This relationship is readily explained by the idea that electrical flow is due to the movement of separate charged particles—electrons—through the metal's atomic structure.

But at temperatures near absolute zero, a metal's electrical behavior is strikingly different. For example, at ordinary temperatures, electricity flows only so long as a driving force from a battery or generator is imposed on the circuit. When the power is cut off, the flow of electricity stops. In a superconducting metal this familiar effect does not occur: an electric current can be made to flow for months after the voltage is cut off.

For a number of years physicists labored to explain superconductivity. Many attempts were made to discover which of the known properties of separate electrons leads to the prediction that superconductivity would occur at low temperatures. None of these efforts succeeded. Success came only when it was realized that in superconducting metals there are no separate electrons but only a kind of collective electrical fluid. Calculations were then developed which showed that although separate independent electrons exist in a metal at ordinary temperatures, at very low temperatures they interact with the metal's orderly atomic structure in such a way as to lose their individual identities and form a coordinated, collective system which gives rise to superconductivity.[5]

No mysterious analog to the "vital force" is at work here, for it can be shown that the property of separate electrons which, at low temperatures, generates the superconducting collective is simply their momentum. This property was well known in separate electrons. But there was no hint in this knowledge that this single property, among the many that characterize the electron, could, when properly interacting with a metal's structure, convert a swarm of separate particles into a correlated whole which exhibits a new quality never perceived in a system of separate electrons.

Such discoveries of modern physics (superfluidity, which causes anomalous collective behavior of atoms in fluids at very low temperatures, is a similar one) show that the unique properties of a complex system are not necessarily explicable solely by the properties that can be observed in its isolated parts. On these quite natural grounds we can expect a similar situation in the marvelously complex chemical system of the living cell. If collective interactions among the separable chemical components of the cell are decisive to the life of the cell, then we cannot hope to discover them by studying only the cell's isolated parts. To comprehend life we need to know its separable constituents, and also to understand that these contribute to life only as dependent parts of an organized whole. If a slogan is needed, it is not "DNA is the secret of life," but "life is the secret of DNA."[6]

The interpretation of the molecular basis of inheritance which I have put forward here is certainly very much a minority view. But it is not a minority of one. For example, George Gaylord Simpson, who is one of the world's most distinguished evolutionary biologists, had this to say recently:

. . . in my opinion nothing that has so far been learned about DNA has helped significantly to understand the nature of man or of any other whole organism. It certainly is necessary for such understanding to examine what is inherited, how it is expressed in the developing individual, how it evolves in populations, and so on. Up to now the triumphs of DNA research have had virtually no effect on our understanding of those subjects. In due course molecular biology will undoubtedly become more firmly connected with the biology of whole organisms and with evolution, and then it will become of greater concern for those more interested in the nature of man than in the nature of molecules.[7]

The Crisis in Biology

There is, I believe, a crisis in biology today. The root of the crisis is the conflict between the two approaches to the theory of life. One approach seeks for the unique capabilities of living things in separable chemical reactions; the other holds that this uniqueness is a property of the whole cell and arises out of the complex interactions of the separable events of cellular chemistry. Neither view has, as yet, been supported by decisive experimental proof. The molecular approach has not succeeded in showing by experiment that the subtly integrated complexity and beautiful precision of the cell's chemistry can be created by adding together its separate components. Nor has the opposite approach, as yet, discovered an integrating mechanism in the living cell which achieves the essential coordination of its numerous separate reactions.

This crisis has deeply affected the course of modern biological research. Since the molecular approach to biology suggests that a complex life process may depend crucially on the chemistry of a specific substance, much of the research has become a hunt for the molecular keys to biological puzzles. Students of the cell have searched for a substance which might dictate the precise rhythm of cell division; embryologists for substances that might determine whether a simple cell develops into a nerve or a muscle; zoologists for molecules that control the metamorphosis of a tadpole into a frog; pathologists for a bacterial substance that causes the invaded host to sicken and die; psychiatrists for biochemical agents that can induce, or cure, schizophrenia.

Some of these efforts have been brilliantly successful; examples are the discovery of hormones such as insulin, antibiotics which combat bacterial growth, vitamins which have a decisive influence on a wide variety of biological processes. But even these familiar successes are marred by illuminating failures. Insulin does indeed repair the major inadequacies in a diabetic, but the progressive disease which persists in many insulin-treated patients indicates that the whole story is much more complicated. Antibiotics do in fact prevent the growth of certain dangerous bacteria, but in practice many bacteria adapt to this insult and develop strains that are resistant to the antibiotic. Vitamins have surely cured pellagra and scurvy, but our understanding of their full role in the body is still so incomplete that the damaging effect of a relatively small overdose of vitamin D is only a very recent discovery. The modern search for new synthetic drugs, which is based on a similar philosophy, has had the same history of limited if important successes and persistent—and too often ignored—failures. Even the successes tell us that there are no absolute molecular controls over biological processes. A particular substance may be especially important in regulating such a process, but the rest of the complex chemistry of the cell is important also. The limitations of what can be accomplished by a single molecular intervention into a biological process testify to the ultimate importance of the whole.

The dominance of the molecular approach in biological research fosters increasing inattention to the natural complexity of biological systems. This has been particularly true of those aspects of biology which cannot be neatly enclosed in a cage and studied in a laboratory—those which constitute the living environment into which all animals, plants, and man must fit or perish. Too often, today, we fail to perceive this system as a complex whole. Too often has this blindness led us to exaggerate our power to control the potent agents which we have let loose in the environment. Only too often in the recent past has our unperceived ignorance led to sudden hazards to life—contamination of our streams with powerful but poorly understood biochemical agents; pollution of the air with powerful but poorly understood radiation.

One of the conceits pressed upon us by the illusory successes of molecular biology is the idea that life is, after all, nothing but a mixture of chemical reactions and that "the boundary between life and non-life has all but disappeared."[8] If we fail to appreciate the profound connection between ignorance and death and persist in our unwitting efforts to disseminate into the environment substances known chiefly for their power to kill, this statement may, after all, turn out to be true.

4

Society *versus* Science

BEHIND every technological innovation, from the undiluted success to the spectacular failure, is the source of the knowledge which technology applies to the solution of a given problem—the free inquiry into nature that we call basic science. I have already pointed out that biology is troubled by a basic theoretical crisis and that this weakness may help to explain why new technological innovations are so frequently troubled by biological failures. Are there other general difficulties in basic science which might also account for some of our technological failures?[1]

Scientific research in the United States gives the appearance of great vigor and health. The recent growth in scientific personnel, in laboratory facilities, and in money spent on research has been phenomenal. Scientific research is the fastest-growing segment in our social structure, its rate of growth exceeding even that of the military establishment.[2] More important is the impressive recent advance in scientific knowledge. In physics: the discovery of fundamental symmetry in the universe, the mastery of increasingly powerful energetic particles, and new insights into the structure of matter and into the origin of the universe. In chemistry: remarkable progress in our understanding of the principles which govern chemical reactions and in the synthesis of increasingly elaborate molecules. In biology: control of once fatal diseases and basic advances in analyzing—if not in interpreting—the chemistry of life.

This is, of course, a very impressive record. But let us also remember that small defects, if unperceived, can produce a catastrophic failure. By applying the traditional skepticism of science to the recent growth of science itself, we may find clues to the surprising failures of some of the new large-scale technological applications.

Traditional Principles of Science versus Modern Realities

Science has well-established methods of work and scientists have long known that the strength of their discipline derives from these principles. But the conditions in which science operates have been changing rapidly and some of the old ideals seem far removed from present realities.

Free dissemination of knowledge has been a traditional principle of science, and the source of its unique capability for self-correction. Knowledge about specific applications of scientific knowledge has long been restricted by military or commercial secrecy. But now even *basic* scientific work is often controlled by military and profit incentives that impose secrecy on the dissemination of fundamental results.

Another of our long-cherished ideals is that science, like the natural world which it seeks to understand, is blind to national boundaries, and that scientific progress is accelerated by the broad sharing of knowledge among scientists of all nations. But some of the massive new research programs—such as exploration of the moon—seem to be intended as competitions for national prestige rather than as contributions to the collective knowledge of man.

The freedom to choose his own problem is the scientist's most precious possession. At the cutting edge of science, on the frontiers of knowledge, nature confronts the scientist with a tangled obscurity which he can hope to penetrate only occasionally and with the most intense and dedicated effort. This kind of effort comes of devotion born of free choice, and scientists have therefore resisted external restraints and blandishments. But recent changes in the pattern of research support, especially the predominance of politically motivated programs such as the space effort, may seriously limit the freedom of choice that scientists now enjoy.

There are serious disparities between the traditional principles of science and modern realities. Perhaps the old ideals are no longer valid, and the present departures from them are a successful adaptation of science to its new position of power and importance. Or are the traditional principles of science still applicable? And in any case, what are the consequences of the growing tendency to encroach upon them? Does this tendency weaken science and impair its technological usefulness; could it cause some of the evils which seem to follow so closely on the heels of modern scientific progress?

Secrecy and Space Science

No one has ever claimed a *scientific* virtue for

military security regulations. The purpose of restraints on free communication about scientific
subjects that relate to military strength is the
frank intention to hamper the development of
comparable knowledge outside the individual nation. These restraints are acknowledged and often
deplored. They are defended only as a lesser evil
demanded by life in an evil world.

Unfortunately, under modern conditions there
are often close ties between military operations
and *basic* scientific research, and the secrecy which
is generated by military demands can work against
the progress of science. Particularly illuminating
examples can be found in our recent experiences
with large-scale space experiments.[8]

One of the main products of post World War II
military work was the development of powerful
ballistic rockets. Although these were designed
primarily to carry nuclear warheads, they are also
able to carry loads of scientific instruments high
above the earth. Data reported back from such
rockets led to the discovery of a wholly unsuspect-
ed feature of our planet—that it carries with it on
its passage through space bands of atomic particles
held in certain zones around the earth by its mag-
netic field.

These newly discovered zones—the Van Allen
belts—interact strongly with radio waves used in
long-range communications, a fact which suggested a particular military application. Government researchers proposed that a nuclear weapon
exploded at high altitude might inject additional
atomic particles into the Van Allen belts and
thereby seriously disrupt radio communication—a
capability of some military importance.

This idea, like many military concepts, was

"born classified" and was discussed only among those personnel who had access to such secret matters. In August 1958, three nuclear bombs were exploded secretly by the United States over the South Atlantic. When this fact was made public some six months later the disclosure was met with a vigorous protest from scientists in the United States and abroad. They complained that so little was as yet known about the newly discovered belts that such gross experimental intervention might make long-lived changes in the natural bands and thereby confuse future scientific studies. This is precisely what happened.

In April 1962, the United States government announced that a new test—Project Starfish—was planned, and again scientists protested that the explosion might produce enough atomic particles to cause large and persistent changes in the natural Van Allen belts. It was argued that such effects might damage experimental satellites, increase the radiation hazard to astronauts, interfere with future radioastronomical observations, and hamper further study of the natural Van Allen belts.

A hot scientific controversy developed over these issues, but because of the secrecy restrictions under which the experiment operated, open discussion of differing views was rather difficult. The United States government asked a group of experts for advice. Meeting under secrecy restrictions, the committee reached the conclusion, which was then made public, that "there is no need for concern regarding any lasting effects on the Van Allen belts and associated phenomena." According to the committee, the explosion's effects on the belts would last "a few weeks to a few months."[4]

On July 9, 1962, still governed by military security, the high-altitude explosion was set off over the Pacific. For some weeks no official information about the results was announced. But despite official silence and the project's military classification, the results of the explosion were soon no more secret than the Van Allen belts themselves. Radioastronomers in Boulder, Colorado, detected on their instruments new radio reflections from the Van Allen belts. These showed that, in the vicinity of the explosion, new zones of atomic particles had been created and that, contrary to the prediction of the government's advisory committee, these were persistent. A check with colleagues in Hawaii and Japan quickly confirmed the results. Thus, by pursuing their own quite open studies, the Colorado astronomers found themselves in possession of information which contradicted government statements about a classified military project. When the information was reported to newspapers, the secret was out. Shortly thereafter the presence of persistent bands as an aftermath of the Starfish explosion was acknowledged by an official government statement.

For months afterward there was an intense scientific debate—partly in the open and partly under security restrictions—about the expected longevity of the new belts. After a year of controversy, a scientific review reported a consensus that in certain parts of the Van Allen belt "it may be necessary to wait more than thirty years"[5] before the natural situation is restored. It was also reported that several experimental satellites, including a joint U.S.—British experiment, *Ariel,* were severely damaged by radiation from the Starfish explosion. The report of the government's advisory committee working under the limitations of secrecy had failed to provide an accurate prediction of the consequences of the high-altitude nuclear explosion.

Secrecy and the Fallout Problem

The fallout problem is another useful illustration of the effects of secrecy on our understanding of large-scale interventions into the environment. Until 1954 basic data on the results of nuclear testing were under stringent secrecy restrictions which, even then, were only partially lifted.[6] Almost no information appeared in scientific journals; most scientists did not know the extent to which fallout had begun to contaminate the environment. Evaluations of fallout contamination had to be made by small groups of "cleared" scientific advisers to the responsible government agencies.

Nuclear contamination involves a wide and complex range of sciences, not only nuclear physics, but also the physics of the stratosphere, meteorology of world air masses, ecology, nutrition, radiation biology, genetics, and pathology. No small group of scientists, however carefully selected, can reflect the full knowledge of the total community of scientists on such a range of subjects. It is not surprising, therefore, that after 1954, when the curtain of secrecy was raised somewhat, exposing the fallout problem to the attention of the entire scientific community, independent scientists made important new discoveries about fallout.[7]

A partial list of such contributions is impressive. The importance of iodine-131 as a hazard to the child's thyroid gland was first suggested by a geneticist, E. B. Lewis, of the California Institute of Technology. The noted chemist Linus Pauling first demonstrated that carbon-14 generated by nuclear explosions is an important biological hazard. Evidence of high local concentrations of fallout in regions near the Nevada test site was first developed by Norman Bauer, a chemist, of Utah State University, and by E. W. Pfeiffer, a zoologist, of the University of Montana. The great value of large-scale analyses of baby teeth as an index of strontium-90 absorption by children was first suggested by a biochemist, Herman Kalckar of Harvard, and the first actual project to collect such teeth, the Baby Tooth Survey, was initiated by the St. Louis Committee for Nuclear Information. The Canadian botanist Eville Gorham first reported the extraordinary capacity of lichens to absorb fallout and indicated the significance of this effect in amplifying the fallout hazard in the Arctic.

The military hold no monopoly on the imposition of scientific secrecy; industrial competition may have the same result. A recent scholarly review of the toxicology of weed killers, written to enlighten the scientific community and to encourage new work on this difficult problem, states in its opening paragraph:

"Many of the toxicological data underlying assessment of the risks involved by using them [weed killers] in practice originate from confidential, non-published reports placed at the disposal of the authorities concerned. Such data have not been included in the present survey."[8]

The High Cost of Secrecy

Problems as intricate, subtle, and pervasive as worldwide contamination from fallout, pollution of the environment by pesticides and herbicides, or the effects of a nuclear explosion on the newly discovered Van Allen belts cannot possibly be solved by committees in secret. Such problems touch on so vast a range of basic scientific questions as to require the full knowledge of the total community of scientists. This community can function only if there is free communication of facts and ideas. Secrecy hinders this essential part of the scientific process.

Scientific knowledge cannot be created whole in one man's mind, or even in the deliberations of a committee, because each separate scientific analysis yields an approximate result and inevitably contains some errors and omissions. Scientists get at the truth by a continuous process of self-correction, which remedies omissions and corrects errors. In this process the key elements are open disclosure of results, their general dissemination in the community of scientists, and the resultant criticism, correction, and verification. Anything that blocks this process will hamper the approach to the truth. The basic difficulty with secrecy in science is that mistakes made in secret will persist.

One of the major costs of secrecy in science has been the serious delay in our understanding of the full consequences of recent large-scale technological operations. Secrecy has deprived us, for example, of the knowledge that might have warned us in time that nuclear explosions are biologically risky, and, when carried out at high altitudes, can obscure for a long time what we want to learn

about the newly discovered bands of atomic particles which surround the earth.

Science for National Prestige

The United States space program is sometimes
regarded as proof of the strength and vitality of
the nation's scientific establishment. Certainly the
spectacular flights of U.S. astronauts into space—
like the similar successes in the U.S.S.R.—are
proof that elaborate achievements in science and
engineering not so long ago regarded as impossible have now become matters of routine. Whatever its strengths and weaknesses, the space program dominates U.S. science; and what we can
learn here from a closer examination may help
explain what is happening on the other fronts of
science.

The nation's space program is a particularly useful source of information about the conflict between the ideal of free international scientific
communication and the modern reality of competition among nations for scientific prestige.
Several distinguished scientists who testified in
hearings before the Senate Committee on
Aeronautical and Space Sciences in June 1963
gave plausible reasons why the space program
should serve the purpose of enhancing national
prestige. According to one of them, we should
regard the race to land on the moon as a kind of
"science Olympics," providing a friendly but impressive exhibition of the superiority of our society to that of the Russians. (Or vice versa if the
outcome is different?) Another witness held that
such scientific competition was needed to demonstrate this nation's technical capabilities in space

and thereby impress upon a possible enemy a realistic appraisal of our strength.[9]

What are the consequences of this position? The exploration of space has been compared, properly, I believe, with its historical analogs—early exploration of the ocean and of distant continents. Explorations of space, like the earlier explorations, are great adventures because they are bold, and they are bold because they are hazardous.

The first visitors to the moon will face extreme dangers, and the scientific success of the program will depend on the safety of the lunar explorers. Will their safety and the scientific usefulness of the expedition be better served by a program driven by the desire for national prestige, or by one based on collaboration with the Russians?

We have had some experience with hazardous extensions of the human habitation. For example, in Antarctica, we—and the Russians—have avoided a competitive search for prestigious accomplishments and work together for the safety and efficiency of both groups. The value of such collaboration in Antarctic exploration, in which we share information, materials, and men with the Russians, is obvious. In a vast, intemperate environment two or three tiny isolated encampments can achieve a common strength much greater than the sum of their separate capabilities. Weather information, if shared, is qualitatively improved for all. Such international collaboration is common on the high seas and in the air: witness the well-established international practices of navigational control and rescue.

If we find it useful to collaborate with the Russians, and others, in such moderately hazardous enterprises, surely in the far greater risks to be

taken on the moon the first explorers should be
given the same advantages. How much more effective *both* American and Russian expeditions
would be on the moon if they were coordinated
and planned for mutual support. How much better scientific information could be obtained by
two parties, in close communication across the
moon's terrain, than if each were there in lonely
and precarious splendor. Sober evaluation of the
realities of this hazardous venture, and the same
sense of responsibility that has guided us in exploration of the earth, weigh heavily against the political advantages of a space program narrowly
impelled by the urge to win a supposed race to
the moon.

The scientific ideal of international collaboration is still valid. Nature is the same everywhere
and what men everywhere have learned about
nature, taken together, leads to knowledge far
deeper than that which each man alone can learn.
It is this collective knowledge that we call science.
Whatever stands in the way of free collaboration
among scientists, whether it is military secrecy or a
competitive struggle for national prestige, weak-
ens science.

Freedom of Choice in Research

Scientists are very jealous of the freedom to
choose their own problems. Of course, much good
science has derived from the solution of an assigned problem and there are many excellent in-
vestigators who advance science through such
work. Nevertheless, the vitality of the total scientific enterprise is strongly dependent on the free
inquiry into nature that we call basic research. In

basic science there is good reason to want a pattern of development which comes from within and to resist the total determination of its course by external demands.

Until recently this principle has, with relatively minor exceptions, been honored by the nation's system of research support. The national Science Foundation was established by Congress with the clear purpose of providing research support broad enough in scope and large enough in amount to encourage the free choice of direction in basic research.

This situation has now been drastically changed. In recent years about one-third of the total federal obligation for basic research—and federal funds are now the major source of support for basic science—comes from a single agency, the National Aeronautics and Space Administration. The NASA budget for basic research has been about six times the amount of the basic research funds administered by the National Science Foundation itself. Unlike the NSF, NASA is a mission-oriented agency, for the enabling legislation requires that it devote its full effort to the conquest of space. Moreover, for the present the mission of NASA is even further narrowed, being centered on a single predominant project—Apollo, which is intended to take a man to the moon and back by 1970. That this single mission has narrowly determined the course of scientific investigations supported by NASA is quite evident from the extensive summary of space research published by the National Academy of Sciences in 1962. The findings of this study state:

"If the Apollo time schedule is to be met, data acquisition necessary to support engineering decision for this mission must take precedence over the acquisition of other data of possibly greater basic scientific interest."[10]

NASA research requires an appreciable fraction of our total scientific personnel. NASA testimony before the Senate Committee on Aeronautical and Space Sciences in November 1963 claimed that space research in 1970 would require the services of only about 6 percent of the nation's scientists and engineers. Yet in 1964 a questionnaire distributed by the American Association for the Advancement of Science showed that 12 percent of a random list of AAAS members received, directly or indirectly, federal research support in connection with space activities. The NASA program is bound to have a considerable directive effect on the over-all course of scientific investigation in the United States.

What is at issue here is that the space program, and therefore its effects on the course of science, has not been created in the interests of *science*. Of course investigations in space involve numerous exciting and important basic scientific questions. But the same can be said of many other fields of science. The problem is to maintain a system of support which reflects the relative significance of each field within the total structure of science. The decision to land a man on the moon by 1970, which in turn dominates NASA-supported basic research, was not dictated by scientific considerations. This fact was clearly stated by Dr. Jerome B. Wiesner, who was President Kennedy's chief scientific adviser when the President made his decision to establish the man-on-the-moon project. At a Congressional hearing, in response to a question as to whether he considered President

Kennedy's decision to establish the project as a national goal a wise one, Dr. Wiesner replied:

"Yes. But many of my colleagues in the scientific community judge it purely on its scientific merit. I think if I were being asked whether this much money should be spent for purely scientific reasons, I would say emphatically 'no.' I think they fail to recognize the deep military implications, the very important political significance of what we are doing and the other important factors that influenced the President when he made his decision."[11]

President Kennedy's political decision to establish the man-on-the-moon project, which he announced in May 1961, appears to have had an important impact on the course of NASA-sponsored basic research. In February 1959, in a comprehensive report on recommended priorities for basic research in space, the Space Science Board of the National Academy of Sciences, which is NASA's chief source of scientific advice, proposed the manned landing on the moon as one project in a list of twenty-one recommended investigations. The report assigned to this project the lowest of three possible priority ratings. Alternative studies of the moon—for example, return of samples of the moon's surface to the earth by means of an unmanned instrument—were given higher priority than the manned exploration of the moon.[12] After the Apollo project was announced by President Kennedy as a national goal these priorities were reversed, and all other NASA research activities became subordinated to the drive to put an American on the moon by 1970.

That scientists may have been reluctant to govern their research interests by acceding to this "national goal" is suggested by the following statement made by a NASA official in an address at the annual meeting of the American Association for the Advancement of Science in December 1962:

"At this stage, the scientific community has the opportunity to assist in determining what man will do out in space, and in particular what he will do when he gets to the moon. If the scientific community does not give this matter its thought and attention and proffer its suggestions and advice, its ideas will be missed. But this will not bring things to a halt. Someone else will make the scientific decisions."[18]

There is good reason for concern that the free pursuit of basic research is being severely restricted by the nation's political goals in space.

The Erosion of Science's Integrity

Some of the recent examples of large-scale experimentation reveal an erosion of the principles which have long given science its remarkable capability to understand nature. These and similar examples were analyzed in detail in a report published in 1965 by the Committee on Science in the Promotion of Human Welfare of the American Association for the Advancement of Science.[14] Some of its findings are worthy of a careful reading:

The ultimate source of the strength of science will not be found in its impressive products or in its powerful instruments. It will be found in the minds of the scientists, and in the system of discourse which scientists have developed in order to describe what they know and to perfect their understanding of

what they have learned. It is these internal factors—
the methods, procedures, and processes which scientists use to discover and to discuss the properties of
the natural world—which have given science its
great success. We shall refer to these processes and
to the organization of science on which they depend
as the *integrity of science*. . . . On the integrity of
science depends our understanding of the enormous
powers which science has placed at the disposal of
society. On this understanding and therefore, ulti-
mately, on the integrity of science, depend the welfare and safety of mankind. . . .

Under the pressure of insistent social demands,
there have been serious erosions in the integrity of
science. . . .

The decision to land a man on the moon was
hardly a fortuitous outcome of the search for knowledge. It was, rather, a decision, largely on political
grounds, consciously to develop the basic and applied science necessary to achieve this particular
technological accomplishment. . . .

Under these conditions, the laboratory of basic
science inevitably loses much of its isolation from
cultural effects, and becomes subject to strong social
demands for particular results. This new relationship has, of course, greatly reduced the delays which
previously intervened between discovery and application. However, the new relationship has also had
a less fortunate effect—*it has resulted in technological application before the related basic scientific
knowledge was sufficiently developed to provide an
adequate understanding of the effects of the new
technology on nature.* . . .

. . . new, large-scale experiments and technological developments of modern science frequently lead
to unanticipated effects. The lifetime of the artificial
belts of radiation established by the Starfish nuclear
explosion was seriously underestimated in the calculations which preceded the experiment. Synthetic

detergents were committed to full-scale economic exploitation before it was discovered that an important fault—resistance to bacterial degradation in sewage systems—would eventually require that they be withdrawn from the market. The hazards of pesticides to animal life were not fully known until pesticides were massively disseminated in the biosphere; the medical risk to man has hardly been evaluated. Nuclear tests responsible for the massive distribution of radioactive debris were conducted for about 10 years before the biological effects of its most hazardous component were recognized . . . .

There is, then, a clear connection between our recent technological mistakes and the erosion of the basic principles of scientific discourse. We seem to be entering a new world of technology, but the vehicle which is carrying us—science—shows dangerous signs of inadequacy for the voyage ahead.

5

The Ultimate Blunder

In the earlier chapters I have tried to show that we tend to use modern large-scale technology before we fully understand its consequences, especially for life. But the specific hazards cited to illustrate this fault are either remote in time (for example, the genetic effects of fallout from nuclear testing, or the possible influence of atmospheric carbon dioxide on the earth's temperature) or seemingly insignificant on the scale of human problems (as, for example, the hazards of insecticides to wildlife). However, one aspect of modern technology is as immediate as this year's tax bill, and vastly more serious than any other issue in the history of man, and that is nuclear war. And here, too, we can find at work the same forces that have led to our blunders with detergents and nuclear fallout.

Nuclear war dominates our lives like an awesome but distant storm cloud. Its horrible face is hidden by military secrecy, confused by technical details; its features are softened by partial truths, wishful thinking, and the banalities of political discourse.

Occasionally some perception of the real nature of nuclear war breaks into the open and reaches the public consciousness. In 1953 President Eisenhower somberly announced that there were grave, hitherto-unmentioned issues at stake in the nation's nuclear war program and promised to share his concern with citizens in a planned-for informational program, "Project Candor." President Kennedy, in an address before the United Nations General Assembly, warned:

"Every man, woman, and child lives under a nuclear sword of Damocles, hanging by the slenderest of threads, capable of being cut at any moment by accident or by miscalculation or by madness. The weapons of war must be abolished before they abolish us."

Former Premier Krushchev spoke similarly:

"In the conflagration of such a nuclear war millions of people would perish; great cities would be razed from the face of the earth; unique cultural monuments created by mankind throughout the ages would be irrevocably destroyed, and vast territories would be poisoned with radioactive fallout."[1]

And Pope John XXIII also warned:

"There is nevertheless reason to fear that the mere continuance of nuclear tests, undertaken with war in mind, can seriously jeopardize various kinds of life on earth."[2]

But even these fitful glimpses into the hell of nuclear war have been dimmed by controversy and confusion. Despite President Eisenhower's urgent hopes, Project Candor never saw the light of day. The Soviet leaders who recognize the inadmissible destructiveness of nuclear war have been violently disputed on this issue by their Chinese comrades. In the 1965 Vatican Council II there was an unresolved struggle between bishops who wished to condemn nuclear wars as a threat to the survival of mankind and others who wished to soften this assertion. Despite President Kennedy's anxiety over the threatened nuclear holocaust, his administration's largest effort at public education on this subject, the distribution of millions of booklets called *Fallout Protection—What to Know and Do About Nuclear Attack,* fell far short of explaining even his own concern. The booklet ignored the fatal inadequacies of fallout shelters for protection against the blast and fire of a nuclear attack; it failed to mention the vast devastation of the nation's productive machinery and agriculture which might make the recommended shelter program an utterly futile labor. And while President Kennedy was warning that nuclear weapons must be abolished, his own Assistant Secretary of Defense for Civil Defense complained:

"The myth, which has been perpetuated by presidents, premiers, and Popes, that nuclear war would be the end of civilization does not stand up under close analysis."[3]

Can we possibly get the facts about nuclear war?[4] Military matters are of course blanketed by secrecy, and the ordinary citizen, or scientist, can hardly expect to be fully informed about them. This is certainly true of offensive preparations for nuclear war. The details of nuclear bomb construction do not appear in scientific journals. No citizen or scientist outside the military ever sees a nuclear bomb; the planes that carry them are only faint sounds in the sky and the missiles aimed at foreign targets lie hidden in underground burrows.

But in one respect, the technological data about nuclear war have come to the surface. A full-scale nuclear war would be directed not at armies and navies, but at people and cities. Defense against nuclear war, although an essential part of a nation's military preparations, is therefore a matter in which the public *must* be involved. In order to develop a civil defense program, the government has published a considerable amount of information about the nature of nuclear war. With this information in hand it is possible to examine at least the defensive aspect of nuclear war technology. And since no military system can succeed in its basic mission, which is to preserve our society, if its defensive efforts fail, the information about civil defense provides a crucial test of the most fateful question of the day: "Can we defend the nation by nuclear war?"

The Weapons of Modern Warfare

Civil defense is concerned with protection against attacks from the nuclear, chemical, and biological weapons that make up the armament of modern large-scale warfare. The most efficient nuclear explosive is the fission-fusion-fission bomb. A small part of this bomb is a trigger, made of uranium 235 or plutonium-239. When the bomb goes off, the uranium or plutonium atoms split, releasing radioactive particles—especially neutrons—and creating fantastic temperatures of about 100 million degrees centigrade. This causes a fusion reaction in the second part of the bomb; atoms of heavy hydrogen fuse together, producing a copious output of new neutrons. These in turn cause fission in an outer jacket made of uranium-238, releasing most of the total power and radioactivity of the bomb.

The development of this bomb in 1954 was an important turning point in nuclear warfare. Until then nuclear weapons were based on the fission of uranium-235 or plutonium-239, or on this process plus fusion of heavy hydrogen. These atomic ingredients are very rare substances, requiring elaborate and expensive purification before they can be used in bombs. In contrast, uranium-238 is much more plentiful and can be made at a cost of less than twenty dollars a pound. The triple-action modern bomb is therefore cheap and essentially unlimited in size.

A modern nuclear war could be fought with huge explosive power. The first nuclear bombs dropped on Hiroshima and Nagasaki released power equivalent to about 20,000 tons of TNT. Each of the new bombs can easily release 10 million tons (10 megatons) of explosive power, and one bomb tested by the U.S.S.R. released over 50 megatons.

The immediate effects of nuclear bombs are quite well known. When the bomb goes off it violently displaces the surrounding air; this blast effect travels through the air, exerting a crushing force on anything in its way. Blast pressure of about 5 pounds per square inch is enough to destroy a brick or wooden house. A 20-megaton bomb would produce such blast destruction in a circle about 22 miles in diameter.

An exploding nuclear bomb puts out a great deal of radiation, some of it as light and heat, and some in the form of X rays and similar high-energy radiation. The radiated heat from a 20-megaton bomb is sufficient to burn all combustible materials in a circle 64 miles in diameter, centered on the point of explosion; this effect is reduced in cloudy weather.[5]

A great deal of the high-energy radiation from a nuclear explosion is not immediate, but delayed. When the bomb explodes, various kinds of radioactive atoms are produced. These are shot into the air by the exploding bomb's fireball; depending on the sizes of the particles to which they are attached, the radioactive atoms will come to earth as fallout at different times after the explosion. The height of the explosion and the weather conditions determine how far the fallout particles will travel before they come to earth. Most will return to earth within a distance of some hundreds of miles from the explosion, but some smaller particles will remain at high altitudes for months, gradually coming down throughout the world as they are caught in rain and snow clouds.[6]

The radioactivity of fallout material decreases with time after the explosion, rapidly at first, and then more gradually. A small (5000-ton) bomb will produce in the nearby downwind regions fallout that emits a total of 12,000 roentgens of radiation; by the second day, three-fourths of the radioactivity is given off and the remainder is emitted gradually thereafter.

A person exposed to about 700 roentgens over the whole body will almost certainly die quickly from radiation sickness. Such a radiation dose might be received almost instantaneously by a person fully exposed to the explosion of a 1-megaton bomb at a distance of about 2 miles and might be received in a matter of minutes or hours from exposure to fallout freshly deposited by the recent explosion of such a bomb. A body dose of about 400 roentgens will kill about half the people exposed to it. Smaller doses will cause radiation sickness during the weeks following exposure and may leave a person much more susceptible to infectious disease. Persons who recover from immediate radiation exposure are more likely than unexposed persons to develop cancer. Domestic animals and plants may also be harmed by radiation. Different species vary widely in their sensitivity to radiation.

We know much less about chemical and biological weapons than we do about nuclear weapons, largely because of more stringent military secrecy. Chemical agents are liquids or gases which, in very small amounts, have powerful effects on living things. The poison gases that have been known since World War I cause sickness and death by damaging the lungs or by burning the skin. The new gases developed since World War II attack the nervous system and are at least ten times as potent as the older ones. They can enter the body through the skin, so that gas masks are ineffective. Some of the new gases cause severe psychological derangements—hallucinations and unpredictable irrational behavior. At higher concentrations, these gases can also kill. They are produced and stocked by the United States, and presumably by other nations as well. Some chemical agents can kill crop plants or remove foliage; these have been used in United Sates air attacks on Vietnam to reduce the protective cover essential in guerrilla warfare or to destroy growing crops.

Biological warfare is the dissemination of highly infectious bacteria and viruses in order to start disease epidemics among people and domestic animals. Some of the known human diseases that could readily be spread in a biological-warfare attack are plague, cholera, and typhoid fever. Some of the animal diseases that could be used in biological warfare are hoof-and-mouth disease, hog cholera, fowl plague, and rinderpest.

Chemical and biological weapons can be sprayed from airplanes or offshore submarines, or spread by special bombs delivered by a plane or missile. In one experiment tiny fog particles capable of carrying infectious agents were spread over nearly 35,000 square miles by a single boat traveling 156 miles along the coast of the United States. Bacteria and viruses can also be spread by releasing insects that have been treated with them. There appear to be no important limits to the availability of chemical and biological weapons. Chemical-warfare agents are readily produced in fairly conventional chemical plants. Biological-warfare agents can be produced in a modified brewery or in antibiotics plants.

Effects of a Nuclear Attack

An analysis of the civil defense problem calls for an estimate of the sizes of possible nuclear attacks on the United States. The United States capability acknowledged by government officials is about 20,-000 megatons of nuclear explosives in a single attack. A corresponding estimate of the nuclear weapons available to our most powerful potential adversary, the U.S.S.R., is more difficult. However, some educated guesses can be made. For example, Defense Secretary McNamara has stated that an attack by the U.S.S.R. on the United States, under present defense and shelter conditions, might kill about 150,000,000 people. An earlier civilian defense analysis by the Rand Corporation,

prepared for the United States government, shows
that about 5,000 to 10,000 megatons of exploding
nuclear weapons would be needed to achieve this
result. An attack delivering this amount of explosive power to the United States would appear
to be within the capability of the U.S.S.R.

To understand the full consequences of an attack, the relative importance of its separate destructive effects (for example, on particular industries) and the ways in which they interact need to
be considered. Some of the particular results of
various kinds of attack are the following:

To kill 60 percent of the United States population
   (about 120,000,000 people) : 500 10-megaton
   bombs dropped on centers of industry.
To destroy 100 percent of the port capacity of the
   United States: 40 10-megaton bombs (400 megatons total) dropped on ports.
To destroy 100 percent of petroleum refinery capacity of the United States: 150 10-megaton
   bombs dropped on petroleum refineries.
To destroy 90 percent of United States heavy industry: 300 10-megaton bombs (3,000 megatons
   total) dropped on centers of such industry.
To burn out about 10,000 square miles of vegetation: 1 10-megaton bomb. Under very "favorable" conditions in a forest area a single 1-mega-
   ton bomb would burn out about 8,000 square
   miles.
To burn all vegetation on 50 percent of the United States land area: an attack with a total of
   about 7,500 megatons of weapons. (Estimates
   are extremely variable; the area that would be
   burned out in a 1,500-megaton attack has been

estimated at from 5 to 80 percent of the total
United States area.)

To destroy by fallout radiation the usefulness of
91 percent of United States cropland and to destroy 95 percent of hog production, 95 percent
of milk production, and 88 percent of all cattle:
an attack totaling 23,000 megatons on combined
military and population targets.

To destroy the usefulness of 44 percent of United
States cropland and to destroy 55 percent of hog
production, 68 percent of milk production, and
46 percent of all cattle: an attack totaling 9,000
megatons on combined military and population
targets.

Taking into account these effects (which assume
that bombs are delivered to different targets with
complete accuracy), the probable inaccuracy of
missiles (which is rather small), and the different
possible patterns of attack, we can arrive at a general estimate of the over-all destruction.[7] An attack of 5,000 to 10,000 megatons would not immediately "wipe out" the nation; some people (about
30 percent) and some food-producing capacity
(something less than 50 percent) would remain. A
heavier attack—at the 23,000-to-25,000-megaton
level which might be possible in the 1970s—would
reduce the surviving proportions of people and
industry to perhaps 10 percent of their former
value. "Total" obliteration, at least as an immediate consequence of attack, would require a considerable increase over the 25,000-megaton level. It
might be achieved by radiological warfare involv-
ing deliberate land contamination with cobalt-60,
or by a 23,000-to-25,000-megaton attack followed
by intensive chemical and biological warfare.

But these figures do not tell the whole story. A major nuclear attack would not only drastically reduce the size of the surviving population but would also markedly change its composition. Because of the uneven geographical distribution of people engaged in various professions, those occupations most important in national recovery would be especially hard hit. Thus a 2,000-megaton attack could kill 73 percent of the nation's architects, 69 percent of the chemists, 62 percent of the physicians, 72 to 86 percent of various kinds of engineers (with the exception of mining engineers), 64 percent of all machinists, 76 percent of tool and die makers.

Could the Population Recover?

A nuclear attack would destroy a massive proportion of the people and economy of the United States. But a society is self-built; given time, a few surviving people might produce a growing population, industry might be rebuilt, and food production restored. A military man might argue that any war has its casualties, and that the basic problem is to determine what casualties we could tolerate and still achieve recovery from the immediate effects of war.

Could the population restore itself after a large nuclear attack? The casualty figures that have already been cited are estimates of the numbers of people that would be killed outright in a nuclear attack by the intense heat and radiation of the fireball, by blast, and by secondary fires, and of those that would die in the ensuing days from the effects of radiation. But the survivors would themselves be threatened by continuing hazards to life.

Many major medical facilities would be destroyed in any significant nuclear attack, since they are concentrated in urban areas which are likely to be targets. For the same reason, a large proportion of the nation's doctors would be killed. Most survivors of a major nuclear attack could not receive adequate medical attention either for the effects of the attack (for example, exposure to radiation) or for ordinary illness. As a result the number of people who would die following a nuclear attack would include many more than those exposed to the immediate death-dealing effects of the attack.

Modern military planners often use a "scenario" to analyze the complex problems of nuclear war by providing a concrete, if imaginary, situation that can at least illustrate what results are to be expected. One such scenario recently prepared by the Hudson Institute, which specializes in war studies for the Department of Defense,[8] is the following:

The United States is struck by an attack totaling 4,000 megatons. As a result of a breakdown in governmental authority border control is inadequate. Several persons engaged in smuggling and illicit activities in the Caribbean are exposed to yellow fever in Haiti, carrying the infection to Puerto Rico and then to Florida. Because of poor medical and police control some of these cases go untreated. Swamp areas such as the Everglades, which support the mosquito carriers of yellow fever, become a reservoir of the disease. A severe epidemic of yellow fever spreads to the Miami area where there are crowded refugee quarters.

According to the scenario: "Mortality of 80 percent among those exposed—80,000 dead in three months." The report points out that similar conditions obtain all along the Gulf Coast and that "some 20 to 30 million people now live in potentially dangerous areas." (It should be recalled that before the development of modern public health measures yellow fever was common in the United States as far north as Boston.)

We are always surrounded by reservoirs of disease; epidemics are held back under modern conditions only by extensive sanitation efforts and by rigorous isolation of potential sources of infection. The destructive effects of any event which disrupts these controls will be enormously amplified by biological reproduction of the disease agents. A nuclear attack may therefore be expected to trigger new outbreaks of diseases largely unknown in the United States in modern times. For example, certain rodent populations in the Southwest harbor fleas which carry a type of bubonic plague. Only the normal isolation of these carriers from concentrated human populations prevents plague epidemics. The incursion of refugees into certain regions of the Southwest to escape nuclear devastation might set off an epidemic of bubonic plague. These are some of the factors that could greatly magnify the initial destructive effects of a nuclear attack.

Such situations, together with the destruction of water supplies and sanitary facilities, inadequate food supplies due to distribution difficulties (the *total* food supply would not be a problem; present stored foods would probably suffice for one to two years), and lack of fuel in cold weather, would certainly increase the amount of illness and raise the death rate. Delayed effects of radiation—increased incidence of cancers, and shortening of life from a variety of immediate causes—would also contribute to an increased death rate among the survivors.

For the same reasons, there would be a reduction in the birth rate. A radiation dosage that can be survived may reduce fertility of both men and women over a number of years. Experience at Hiroshima shows that fetuses exposed to appreciable radiation before birth had an increased frequency of serious defects; the incidence of still births was also increased. Radiation exposure of the population would increase the frequency of harmful genetic changes in later generations and so contribute to the general decline in the population's biological capability.

The ultimate biological fate of the population that might survive a nuclear war would depend on the balance between birth rate and death rate; if the death rate exceeded the rate of biologically capable births, the population would in time die off. If we add the effects of a possible further attack with disease-producing biological weapons, the probability of the population's biological death becomes much greater. These are only generalizations; the problem is too complex and the data are too scarce to permit more detailed predictions.

Even if biologically capable of self-renewal, the population would be able to increase only if the remaining industrial facilities were sufficient to support recovery. The several capabilities which are essential for recovery are mutually dependent. People must have a mimimum amount of food, clothing, shelter, and emergency care, or they will die; in turn, human labor is necessary to produce the means of subsistence. A small change in either capability—for example, an epidemic which reduced labor power, or one which destroyed a food crop—could create a vicious circle which would cause the situation to deteriorate rapidly. The reverse is also true. A sudden increase in available food (say, by the restoration of an important link in the transportation system)' might significantly improve the strength of the labor force and thereby permit increased production of better food, thus starting the social system on a rapid upward trend.

Any society that survived a nuclear attack would therefore be delicately balanced between recovery and further deterioration. Unless it could begin to recover shortly after the attack, such a surviving society would tend to go rapidly downhill. This means that the fate of the society would depend considerably on the *interactions* among the controlling factors. For example, if an attack left many survivors but destroyed most of the facilities needed to sustain life, competition among the survivors could easily set off a cycle of deterioration. Conflict—or even competition—for scarce food would lead to inefficient use of available supplies; recovery efforts would be hindered; shortages would worsen; conflict would intensify; the social system would start on a descending spiral toward disintegration.

Agricultural Devastation

Even if the survival of a human population capable of self-propagation and of sufficient industrial capacity to provide the necessary subsistence could

be assured, the regeneration of society in the
United States after a major nuclear attack would
still depend on success in a third sector—the biology of the earth's surface. No society can survive
without food. Once the one-to-two-year store had
been used up, food would have to be produced
year by year. Agriculture is therefore of crucial
importance in determining whether the surviving
population would be able to restore a functioning
society. According to a Stanford Research Institute
report prepared for the government,[9] an attack of
about 10,000 megatons would cut agricultural
production approximately in half because of the
killing effects of fallout on livestock and contamination of the soil by strontium-90. However, other
estimates show that such an attack would also
burn off the vegetation on more than half of the
United States land surface. When this hazard is
added to the effects of fallout, it appears that a
10,000-megaton attack would reduce agricultural
productivity well below the 50-percent point.
With an attack of 23,000 megatons, which might
be possible in 1970, agricultural productivity
would fall to about 10 percent of its original value.

Some civil defense proponents point out that
grass and crop plants are relatively resistant to
radiation and that it might be possible to obtain
crops—albeit radioactive ones—from fallout-con-
taminated land. Rather than starve, survivors
of a nuclear war could manage on such food, even
though they would be absorbing amounts of radioactivity that would shorten their lives and increase the risk of genetic damage to their offspring.

But fallout can have a much more disastrous
effect on agriculture indirectly—for example,
through its effect on birds. This is succinctly de-

scribed in another Hudson Institute scenario. In this instance it is supposed that the United States is attacked with a total of 5,000 megatons of nuclear weapons, some of which are aimed at knocking out the complex of Minuteman missiles in north-central Missouri. The wind pattern carries heavy fallout into the area of Champaign–Urbana, Illinois. Although, according to the scenario, there would be enough warning time to permit the inhabitants of Champaign County to take shelter against fallout, the entire farming area would be heavily irradiated. The fallout radiation doses would be so high that 80 percent of all adult birds and nearly all of the young ones would be killed within three weeks. The fallout level would also be too high to permit effective farm work, although the crop plants are not themselves damaged by the radiation. As a result of the loss of the bird population and of inadequate farming, within a year the population of destructive insects would increase so much that nearly half of the region's crops would be lost.

The scientific basis for this scenario is well established. Stomach analyses reveal that field birds eat huge numbers of insects; birds have frequently saved crops from destruction by insect invasions. Radiation tests show that birds are killed by doses in the range of 500 to 1,000 roentgens, whereas adult insects are killed only by doses at least ten times as high. On the other hand, at certain stages of development insects are about as sensitive to radiation as birds are. If an attack happened to strike an agricultural area at the season at which most insects were in this sensitive stage of development, there might be no subsequent infestation of the crops. But we have no way of knowing

how the timing and pattern of a nuclear attack on
the United States would fit in with the timing and
geographic pattern of the life histories of the insects that inhabit our fields.

It seems clear that a large-scale nuclear attack
would sharply reduce the food-producing capability of the nation, so that for many years it could
not support a population approaching the present
one. If enough people survived the attack, the
imbalance between available food production and
demand for food might precipitate a cycle of social
disorganization.

The Assault on the Biosphere

Despite the organization imposed on it by human
activity—seeds planted in rows, fertilized, weeded,
and protected from insects—agriculture remains a
part of the larger, over-all system of life which
occupies a thin layer on the surface of the earth:
the biosphere. The success of agriculture depends
on the proper functioning of the biosphere.
Recovery from a nuclear attack requires not only
that agriculture retain its capability to support
human society but also that the biosphere remain
capable of supporting agriculture.

The biosphere is a marvelously intricate system
that ties together the lives of many millions of
varieties of plants, animals, and microorganisms;
the daily and seasonal changes in light and
darkness; the myriad movements of wind and
water that make up the weather; the seasonal
changes in the growth of living things and in the
physical environment; the gradual change in the
chemistry of the earth's surface over the years; and

the ceaseless evolutionary changes in the character of the organisms that inhabit it.

As indicated earlier, all the parts of this system are mutually interdependent; a change in one sector will affect the whole. Forest destruction would be particularly important in a nuclear attack. Pine trees on a mountainside need water and soil for their growth; as they grow, the trees produce a network of roots and a blanket of needles which bind the soil in place. Thus the trees depend on water and soil, and the existence of the soil and its retention of water depend on the trees. A temporary upset in this balance can have a huge and long-lasting effect. If pine trees are killed, the rain will no longer soak into the soil but will wash it away. The mountainside, now bare of soil and unable to retain water, will no longer support the growth of trees. New seeds will not take hold, and the forest will not return to the mountain.

For these reasons, the widespread fires and radiation resulting from a nuclear attack would cause severe changes in the balanced relations of the biosphere, which could have unpredictable but possibly catastrophic effects on agriculture and on the general capability of the land surface to support a successful recovery. The pine tree is unusually sensitive to radiation; the radiation dose required to stop the growth of a pine tree is about 30 roentgens per day during an entire growing season. This is one hundredth of the radiation dose required to stop the growth of crabgrass. Following an attack in the 10,000-megaton range, probably more than half the surface of the United States would receive fallout yielding a total dose of thousands of roentgens. Many pine forests would probably be killed off, which in turn would upset the rain-soil balance, so that the areas could easily become permanently eroded. Large-scale loss of trees can lead to important changes in weather and water conditions in adjacent farming areas, so that the end result may be a serious loss in agricultural capability.

Trees that are more resistant to radiation than pines would not die off so readily, but radiation would make them more susceptible to insect attack. Since adult insects are highly resistant to radiation and their natural enemies, birds, are relatively sensitive, fallout is likely to cause an increase in insect populations. Thus, the ravages of insects in forests may become greatly increased, and the resulting swarms of insects could then move on to attack certain crop plants.

Finally, all human activities, from agriculture to industry, depend on a suitable climate. That this requirement of a successful habitation of the earth is also threatened by a nuclear war is suggested by another scenario from the Hudson Institute. It concerns a type of attack of great severity but one which is nevertheless within the anticipated capability of present nuclear armaments: a protracted war lasting three years in which the United States is attacked by 20,000 megatons of nuclear weapons mainly exploded on the ground to destroy underground installations of missiles and other military systems. Apart from the vast general devastation that such an attack would cause, the scenario points out that it might readily induce a permanent change in the weather of the continental United States that would bring on a new ice age.

The argument developed by the scenario is as follows: The vast quantities of dust injected into the stratosphere by this attack would shade the earth from sunlight and cut down the solar heat received at the earth's surface by 10 percent in each of the three years of the war. According to the scenario, this would set off the following chain of events:

In the fourth winter [after the start of the nuclear war], average temperatures in the northern hemisphere are already 4°C below normal and the snow line of the Sierra Nevada and the Rockies is one-half mile lower than previously. . . . Tampa now has the climate New York formerly had and Seattle has the climate formerly associated with Juneau, Alaska. . . . The precipitation in Washington and Oregon is mostly snow for one-third of the year. Where the snow remains on the ground during the summer still more heat is lost by reflection, and the snowpack at higher elevations continues to deepen and encroach on the valleys. Small glaciers are formed, merge, and grow. Permanent snowcaps are seen on many peaks in the Appalachians as far south as Tennessee. Even at low altitudes snow is found on forested northern slopes in New Jersey until June. North of 50° latitude, where the earth's surface is about 60% land, and much of the rest—the Arctic Ocean—has year-round pack ice, the reflectivity (albedo) of the earth's surface shifts upward by many percentage points as vast stretches of Canada, Alaska, and Siberia are covered by snow for longer and longer times. The result is that cold, polar, high-pressure air masses are more prevalent and move farther south, even in summer. Thus the process continues even after the dust is scavenged from the stratosphere. A period of glaciation in the northern hemisphere begins.[8]

The scenario states: "Nobody would be surprised . . . if a glacial era *once begun* lasted ten

thousand years." As to countermeasures the report
is understandably modest: " . . . the events described in the foregoing are sufficiently unlikely,
yet on such a large scale that it is difficult to
discuss feasible countermeasures."

These are some of the issues that would determine whether the nation could recover from a
nuclear attack or descend toward chaos. There
would be a huge initial destruction—of people,
industry, resources, and agricultural production.
Because of complex feedback relationships huge
effects may be triggered by small, indirect, and
uncontrollable events; although food plants are
not especially sensitive to the damaging effects of
radiation, fallout does, in fact, constitute a major
threat to crops but an indirect one, caused by the
high radiation sensitivity of birds and pine trees.
Vastly different consequences can result from the
different ways in which an enemy chooses to, or
happens to, carry out the attack; catastrophic and
long-term climatic changes might be triggered by
an extremely heavy nuclear attack, but only if the
bombs were exploded on the ground.

Could Civil Defense Work?

In the face of these problems what might be accomplished by any conceivable civil defense program? Fallout shelters would protect some people—
provided that they were not in a direct-attack area
and that no chemical or biological weapons were
used. Stronger shelters might also protect against
the devastating blasts and offer some protection
from fire in areas near an explosion, but the survivors might emerge from the shelters only to be
wiped out by subsequent epidemics. Shelters

would nevertheless save some lives, but if key industries were not also protected there would not be enough of the necessities of life to support the survivors. The very protection of the population might then overwhelm the remaining resources, so that social disintegration and chaos would result. And even if it were possible to provide equal protection for people and industry, there is no way to protect agriculture, the biological balance of the land, and the climate from the effects of a nuclear attack—from catastrophes which could render all efforts to protect the population and industry completely futile.

Can we reach any conclusions about whether or not the nation could survive a nuclear war? Obviously this is a matter of judgment. My own conclusion is based chiefly on the numerous possibilities that the huge initial devastation would be enormously amplified by a series of biological processes—epidemics, crop destruction by radiation-induced ecological imbalance, erosion and sterilization of the land resulting from massive destruction of vegetation—and by the triggering of possibly catastrophic climatic changes. The staggering size of the immediate destruction by a large-scale nuclear attack, the probability of great amplification by biological and climatic processes, and the self-contradictory consequences of civil defense measures lead me to conclude that, despite any conceivable civil defense program, this nation, its population, its economic wealth, its social fabric—all that we speak of as our civilization—would be lost irretrievably after a major nuclear war.

Can this conclusion be proven to the satisfaction of all informed observers? Obviously not; other scientists have come to a contrary conclusion.

What, then, can the scientific community say, collectively, to those who must plan the nation's defenses, and to the American people, who must, after all, judge the wisdom of these defenses? Here I *can* speak with certainty, for the commands of our discipline are clear. Although in my judgment a major nuclear war, despite any plausible defense measures, would irretrievably destroy this nation, because I am subject to the discipline of science I am obliged to add that this judgment is not now provable and that others possessing the same information disagree. Obviously the same obligation lies on those who disagree with my conclusion.

Taking into account this difference of opinion, scientists, as a collective community of informed specialists, are obliged to report in answer to the question "Could the nation survive a nuclear war?": "We do not know!"

If, as most would agree, the purpose of the nation's military defenses is to preserve our "way of life," then we cannot now decide on the basis of the available scientific evidence whether it can possibly serve this purpose. At the risk of the very existence of the nation, we harnessed the explosive power of nuclear reactions to the nation's system of defense before we understood that it stands a good chance of failing.

A Huge and Costly Gamble

No one who is not privy to the highest military secrets can know when, in the course of developing the nation's nuclear war machine, our military and political leaders discovered that it is a huge and costly gamble. Perhaps President

Eisenhower planned Project Candor in order to share with the nation his concern on learning the frightening truth about our nuclear "defenses." But it seems clear that the nuclear military system was well developed before military planners were aware of the problem which creates the greatest uncertainty about nuclear "defense"—the ecological effects on the environment.

A 1961 Rand report prepared for the U.S. Air Force on the ecological problems of nuclear war states on its opening page:

> This paper is a first approach to the "Civil Defense Problem" and post-attack recovery and is written from a broad ecological point of view. It is a point of view which has been strangely neglected (although many have been vaguely concerned), and detailed research is conspicuously absent. . . . Many of the ecological principles underlying the problems involved are not part of the intellectual equipment of people ordinarily concerned with Civil Defense and postwar recovery.[10]

When Norbert Wiener warned in 1960 that we might lose control of the complicated machines and systems of machines that modern science has created, he clearly had in mind the most destructive system which man has devised—nuclear war. He predicted that machines could be built that would automatically absorb information about the position and speed of enemy missiles, determine their origin, calculate their apparent target from minute to minute, and then, following built-in instructions, throw the switches to send off retaliatory weapons.

The war machine that Norbert Wiener feared may soon exist in the United States. Modern missiles make automatic control essential. The time between the launching and the arrival of an intercontinental missile would be about 30 minutes or less; without the help of fast calculators, no one could decide what to do about a suspected attack before the bombs arrived. Until a few years ago our nuclear rockets were of the type propelled by liquid fuels. These were large, and slow to activate, and therefore not readily incorporated into an automatic system. But these missiles have now been replaced by the Minuteman—a smaller, solid-fuel rocket which can easily be shot out of its underground burrow in response to an electronic signal.

Military secrecy prevents us from knowing whether the thousand or so Minutemen now in place in the United States are tied into a computer system, how their targets are selected, and how the button would be pushed. But it has already been announced that a computer-operated system is being devised that can track approaching rockets, determine their paths and prospective targets, and automatically send off our own antimissile missiles to intercept, and—it is hoped—destroy the oncoming enemy missiles. It would be a relatively simple matter to use this system to work backward from an enemy rocket's calculated path and determine not only its target but its point of origin. With this information in hand the war computer could readily instruct the waiting Minuteman system as to *its* targets and—with or without an intervening human decision—send them on their way. It would be rather astonishing if this capability were not put to use by our military technicians, creating thereby the full automatic war machine that Norbert Wiener predicted six years ago.

But Dr. Wiener also warned us that such machines are not to be trusted. He suggested that if we build a machine and instruct it to win a war it may, like the New England power grid, do its duty in a completely unexpected way, and ". . . win a nominal victory on points at the cost of every interest we have at heart, even that of national survival."[11]

The data of civil defense show us that the whole modern war machine is larger and much more complicated than even computer-operated missile systems. It includes the vast arsenal of modern military technology: conventional, nuclear, chemical, and biological weapons, as well as high-speed electronic systems for detection and computation, civil defense facilities, the economic and social structure of the nation, and the total environment in which we live.

We have tied the complex fabric of our total civilization into an enormously powerful, lightly triggered, and poorly understood war machine. All the features of the runaway machine that Wiener feared are present. If it is set in motion, it may carry us, unwittingly, unwillingly, to a catastrophic fate. Like some of the other large-scale technologies that are based on modern science, the technology of nuclear war was adopted before we were aware of its ultimate futility.

Like detergents, synthetic insecticides, and leaded gasoline, the weapons of nuclear war demonstrate our mastery of physics, chemistry, and engineering. And like these lesser technological innovations, the nuclear war system is likely to fail in its purpose—defense of the nation—because of unanticipated hazards to life.

Again like the examples cited earlier, the technological blunder of nuclear war has its roots in the breakdown of traditional scientific procedures. Secrecy for a long time protected the mistakes of nuclear military technicians from the scrutiny of the scientific community, and it still hampers open criticism. Had the nature of nuclear warfare been fully described in the scientific literature, biologists could have explained to the generals—long before 1961—that they could not guarantee that the delicate balance of life-relationships on the earth's surface would survive a major nuclear war.

In the field of nuclear weaponry, international cooperation, a principle of scientific discourse which has done so much to enhance the progress of science, was of course an immediate casualty of the cold war. Nevertheless, the few instances, chiefly relating to fallout, in which data have been openly available to the international scientific community demonstrate the advantages of such communication for a true understanding of the problem. Attention was first drawn to the biological hazards of carbon-14 from nuclear explosions by Russian scientists; the analysis was further developed and refined in the United States by Linus Pauling and later confirmed by AEC scientists. Conversely, Soviet estimates of the fallout hazard in the vast Arctic regions of the U.S.S.R. must be greatly aided by the data on the problem in Alaska produced by United States and Canadian scientists. If international exchange of information, however unplanned, was useful in these relatively minor matters, is it not possible that an international convocation of military scientists might have concluded, at a mercifully early date in the nuclear arms race, that the whole system was unworkable? Nuclear war is another—and the most appalling—example of the growing dangers to survival that result from the erosion of the integrity of science.

Our military experts and political leaders must by now be well aware of the risks which this nation takes in relying on nuclear weapons for defense. Each must search his conscience before deciding that there are benefits sufficient to counter-balance such fearful risks. But this is not the kind of judgment which can be left to experts, or even to elected leaders. A decision which may risk the very life of the nation must be made by the nation as a whole. Perhaps the nation is willing to take this risk. But we won't know until it is asked.

6

The Scientist and the Citizen

THIS account has thus far been limited to matters of science and technology, and most of what I have said can be supported or disputed by marshaling appropriate data, either scientific or historical. Thus far, however, I have described problems but no solutions. Since it is impossible to speak of what might be done to alleviate environmental pollution or to save us from the frightful catastrophe of nuclear war without rendering opinions and expressing personal convictions, we must now move from the realm of science and technology into these less certain grounds.

That we *have* learned a little about how to cope with some of the problems created by the new technologies should be evident from the fact that the world has managed to survive the discovery of nuclear power for more than twenty years. We can examine what has been learned, first to find how scientific and technological knowledge can help solve these problems, and then to discover where science stops and public morality takes over.

Mastering Fallout: The Scientific Basis

So long as information about fallout was blanketed under military security, the problem of eval-

uating its effects and what might be done to min-
imize them was necessarily in the hands of the
military, the AEC, and their scientific advisory
groups. During that period, the public was assured that there were no hazards from nuclear tests
and that no protective action was needed. When
secrecy about fallout was lifted, beginning in
1954, the general scientific community became
aware of serious inadequacies in the published
accounts of the fallout problem and developed a
concern about the possible hazards. The next few
years were marked by technical (and other) disputes between scientists associated with the government's nuclear test program and the larger
group of unaffiliated scientists. Most of these
scientific controversies have been resolved, many
of them rather rapidly. For example, when Linus
Pauling first suggested that carbon-14 from nuclear explosions contributed significantly to the
total biological hazard, he was vigorously disputed
by government radiation experts. But as soon as a
government laboratory made a detailed check of
Pauling's calculations, the controversy was settled,
for their answers were in substantial agreement
with his.

Controversy is nothing new to science; it is common when the available data are insufficient to
decide between conflicting points of view. The
remedy is more data, and the fallout controversies
were useful because they revealed the need for
more information about the distribution and
effects of fallout radioactivity. When the previously secret AEC reports were made available to
scientists, it became apparent that fallout data
gathering was seriously incomplete. Sampling was
fairly thorough in the areas near nuclear test sites

but very spotty elsewhere in the United States and in the world. Measurements of radiation in foods were scanty. Only a few of the nation's hundreds of milksheds were sampled for strontium-90 at regular intervals; measurements of other foods were scattered and irregular. There were very few measurements of fallout radiation in surface waters. Almost no data were available on certain fallout constituents, especially strontium-89 and iodine-131, the latter omission being seen later as a serious hindrance to understanding the fallout hazard near the Nevada test site.

Many of these inadequacies were pointed out by scientists during the early fallout controversies. It is to the credit of the United States Public Health Service that it responded vigorously to this situation. Beginning in 1957 the USPHS established a program of fallout monitoring which has grown into an elaborate and widespread system of measurements, taken at frequent intervals, of radioactivity in the atmosphere, in surface waters, in the soil, and in foods. The monthly bulletin which the USPHS now issues on these measurements and on data provided by other United States and international agencies is the most detailed information now available about any aspect of environmental contamination.

These new data-gathering systems have strikingly improved our knowledge of fallout, and previous controversies have begun to give way to fact. The initial controversy over the possible hazard from iodine-131 was largely due to the infrequency of the necessary measurements. Because of its short half-life (half of its initial radioactivity decays in only 8 days) iodine-131 can be detected only if measurements are frequent and detailed.

When the USPHS monitoring system went into operation it became apparent that each nuclear test in the atmosphere was accompanied by a brief but intense introduction of iodine-131 radioactivity into the food chain. Direct measurements of radioiodine in milk gave fairly precise calculations of the resultant exposure, especially to children. It then became evident, for the first time, that in the continental United States iodine-131 is responsible for the most intense human exposure to radioactivity from fallout during testing. More important, it became possible to warn milk producers of the hazard and to devise relatively simple countermeasures, such as temporary diversion of milk supplies from the market, to bring this hazard under control.

The successful interpretation of such monitoring data was possible only because the source of the contamination was clearly established. Every nuclear test in the atmosphere, by the United States and other nations, is recorded as to size, time, and place. United States laws also require reporting of other possible sources of radioactive contamination, such as nuclear reactors. Such a detailed registry of sources, combined with intensive and widespread monitoring, tells us a great deal about how radioactivity spreads into the environment, and how its hazards can be minimized.

Nevertheless, even with these improvements controversies about fallout persisted. These were centered in the establishment of standards of acceptable exposure. When the fallout issue first arose, the only existing radiation standards were those designed for industrial protection, which were not immediately applicable to situations in which whole continents and entire populations were exposed. For this reason, and because the levels of radioactivity due to fallout were very much lower than those encountered in industry, there was no agreement on what standards ought to apply.

A particularly important lack was the absence of a clearcut theory of biological radiation damage. One theory suggested that repair processes might occur, thus protecting the tissues from any permanent damaging effects from exposure to very low levels of radiation. This approach leads to the concept of a threshold dose which must be exceeded if any biological damage is to ensue. In this case a standard of exposure is fairly easily devised simply by setting it below the threshold dose, which can be determined from experiments with laboratory animals and from observations of accidentally irradiated people. In contrast, another concept of radiation damage—the linear theory—held that there was no threshold and that any increment in radiation exposure would proportionally increase the risk of biological damage. In this case there is no absolute way to establish a standard of tolerable exposure. Any exposure must then be regarded as harmful to some degree.

The scientific community has played a decisive role in resolving this conflict. Largely in response to the fallout problem, geneticists carried out elaborate experiments to study hereditary effects at the low radiation levels which approximate those encountered in fallout. Radiation pathologists also pressed their experiments on tissue damage to lower radiation limits. As a result there is now a rather common agreement that the linear theory of radiation damage is the most reasonable guide to radiation standards. The standards adopted by the responsible United States agency, the Federal Radiation Council, are based on this conclusion.

If any increase in radiation exposure, however slight, is accompanied by a comparable increase in the risk of medically undesirable effects and there is no "safe level" of radiation, how can one determine what dosage is to be tolerated? This judgment requires a balance between the risk associated with a given dosage and some possibly countervailing benefit. The Federal Radiation Council explicitly adopted this position in 1960:

> If . . . beneficial uses were fully exploited without regard to radiation protection, the resulting biological risk might well be considered too great. Reducing the risk to zero would virtually eliminate any radiation use, and result in the loss of all possible benefits. It is therefore necessary to strike some balance between maximum use and zero risk. In establishing radiation protection standards, the balancing of risk and benefit is a decision involving medical, social, economic, political and other factors. Such a balance cannot be made on the basis of a precise mathematical formula but must be a matter of informed judgment.[1]

However, in the actual application of these standards to the fallout problem there has been considerable confusion. When, as a result of atmospheric nuclear tests by the United States and the U.S.S.R. in 1962, the amount of iodine-131 in the milk supplies of several states approached the level which, according to the Federal Radiation Council, required preventive action, local health officials took what they believed to be appropriate measures. In Utah, Wisconsin, and Minnesota, the

state departments of health asked farmers to
divert fresh milk from the market, so that there
would be time for the iodine-131 to decay to acceptable levels before the milk was used for human consumption. But this action was opposed by
the United States Secretary of Health, Education
and Welfare (who also serves as chairman of the
Federal Radiation Council). He stated that the
Council's radiation-exposure standards were applicable only to "normal peace time conditions," and
according to him these conditions did not include
nuclear testing. This interpretation meant that
nuclear testing constituted an additional factor
not included among the "social, economic, political and other factors" which entered into the
Council's original calculation. The Secretary's action can only mean that, in his view, the value of
nuclear testing to the nation warranted some increase in the acceptable medical risk from iodine-
131. This is a clear illustration of how the seemingly technical questions of environmental pollution
very quickly extend beyond the realm of science.

Mastering Pesticides: The Scientific Basis

Quite similar problems are associated with other
forms of environmental pollution, of which the
extensive killing of fish near the mouth of the
Mississippi River is an example. Beginning in
1957, sugar-cane and cotton farmers in the Mississippi Valley began to spray their crops with a
new pesticide, endrin, which is a chemical relative
of DDT. Several years later large numbers of
dead fish began to appear near the mouth of the
river, and set off an intensive investigation and
considerable controversy.

Despite deep-seated disagreement between the disputing parties, two facts are clear and acknowledged: Many fish have died and the Mississippi River contains detectable quantities of several insecticides and related organic compounds. The issue is whether the insecticides are the cause of the fish kills and are a hazard to human health, and if so, what should be done about it.

An important argument centers around the possible harm to humans from the very small amounts of pesticide residues present in edible fish and in drinking water taken from the river. One side points to the fact that some laboratory animals exhibit no toxic effects unless exposed to pesticide concentrations many times greater than those due to river pollution. The other side points out that we have inadequate data on the effects of such small concentrations of pesticides on laboratory animals exposed for long periods of time. The possible effects of chronic, long-term exposure of humans to low concentrations of pesticides are also unknown, because, in a sense, the necessary experiment has only just begun.

This issue is the same one that arose in connection with the fallout problem ten years ago, and the same solution is indicated. Estimates of the hazard must be based on the assumption that any increase in exposure results in a proportional risk to the total living population of the biosphere. Like radiation, many of the new synthetic substances act on basic biochemical processes that occur in some form in all living things; therefore some effect on all forms of life must be anticipated. Since some of these pollutants appear to increase the incidence of cancer and the rate of mutation, it is entirely possible that, like radiation, they act on the cell's system of inheritance. Such changes in inheritance may persist in the offspring of the affected organism. The changes are thereby perpetuated and result in an additive risk of eventual biological harm. Moreover, since the biological system exposed to the pesticides is very large and complex, the probability that any increase in contamination will lead to a new point of attack somewhere in this intricate system cannot be ignored. Finally, because the toxic effects of many organic pollutants, like those of radiation, may appear only after a delay of many years, extreme caution ought to be the rule in the early uses of pesticides and other novel substances that contaminate the environment.

The very presence in the Mississippi River of substances known to be toxic to fish at low concentrations and to mammals at higher concentrations must be regarded as a definite risk to any biological population exposed to it. The only feasible way to judge the significance of this contamination is to estimate the risks, compare them with the benefits associated with the use of the pesticides, and strike a balance between risk and benefit that will be acceptable to the public. This means that we must know the sources of the contaminants and determine, for example, whether the operation that causes their appearance in the river water is the spraying of corn and cotton crops in the river valley or the activity of riverside plants which manufacture pesticides. This, in turn, will require us to adopt the practice of registering not only the manufacture but also the distribution and use of large amounts of pesticides. Without such a registry there may be no way to determine the source of the contaminants. Finally, against the benefits involved must be balanced the economic losses to the river fisheries and the possible but still unknown hazards to the health of people who absorb the pesticides by eating fish or drinking water taken from the river.

A similar approach is, I believe, equally applicable to most other pollution problems. Since they are all large-scale effects and influence a wide variety of living organisms, on statistical grounds alone it is probable that the smallest detectable pollutant level represents some hazard, however slight, and that the risk will increase roughly with the level. Until known risks can be balanced against specific benefits, no meaningful action is possible. But in the absence of such action the rule of prudence, which is demanded by the unknown long-term hazards, requires that extreme caution be exercised in continued use of these agents.

Risk versus Benefit

It appears, then, that the problems of environmental pollution require a common approach: the principle of balancing risk against benefit. The risk can be determined by estimating the number of people exposed to the pollutant, the amounts which they may be expected to absorb, and the physical harm that might result. The benefit can be determined by estimating the economic, political, or social gains expected from the operation which produces the pollutant and the possibilities of substituting less hazardous operations.

Estimations of risk and benefit are proper subjects of scientific and technological analysis. There are scientific means for estimating how many cases of leukemia and of serious congenital defects may result from fallout radiation; such calculations have been reported in great detail by the United Nations Scientific Committee on the Effects of Atomic Radiation. Medical statistics can provide a similar estimate of the amount of respiratory disease that is related to exposure to smog. It should be possible, eventually, to determine what biological risks to humans, birds, or fish are to be expected from a given dosage of DDT.

Determination of the corresponding benefits is more difficult but nevertheless is also within the realm of technological competence. For example, if automobiles are, as they appear to be, a major source of smog in urban areas, it should be possible to evaluate their economic and social importance and to compare it with alternative forms of transport, such as electric trains, which are not smog producers. The economic value of insecticides to the farmer is readily calculable. While no money value can be placed on the benefits to be derived from the development of a new nuclear weapon, it should be possible, it would seem, to determine the necessity of such weapons to the nation, and the importance of nuclear tests to their development.

However difficult the procedures and uncertain the results, all these questions are subject to objective scientific and technological analysis. Presumably scientists who differ in their personal attitudes toward nuclear tests, superhighways, or songbirds could agree, more or less, in their estimates of the relevant benefits and of the associated hazards of fallout, smog, or DDT.

Once the hazards and benefits of new technological innovations become clear, it may be possible to find means to reduce the hazards—at a price. Automobile exhaust emissions can be partially reduced by mechanical devices, which will be required by law on all 1968 models. If we are threatened by accumulating carbon dioxide in the air, engineers can build devices, however expensive, to remove this substance from flue gases. If chemical pesticides are an unwarranted hazard to wildlife or man, we can, after all, stop using them and suffer the sting of the mosquito and the depredations of insects on our crops, while we try to learn enough about environmental biology to develop more natural means of control. If we find strontium-90 intolerable, the nuclear tests that produce fallout can be stopped, and they have in fact been sharply reduced by the test-ban treaty. What is needed is not only the development of technical means for dealing with environmental pollution, but also the willingness to undertake the extra expense and additional inconvenience to prevent the intrusion of pollutants into the environment.

Beyond the Realm of Science

With the determination of benefits and risk and the development of techniques which improve the balance between them, the applicability of scientific procedure to the problems of environmental contamination comes to an abrupt end. What then remains is a judgment which balances the stated risks against the corresponding benefits. A scientific analysis can perhaps tell us that every nuclear test will probably cause a given number of congenitally deformed births, but no scientific procedures can choose the balance point and tell us how many defective births we ought to tolerate for the sake of a new nuclear weapon.

What is the "importance" of fallout, determined scientifically? Some scientists have stated, with the full dignity of their professional pre-eminence, that the fallout hazard, while not zero, is "trivial." Nevertheless I have seen a minister, upon learning for the first time that acts deliberately performed by his own nation were possibly endangering a few lives in distant lands and a future time, become so incensed at this violation of the biblical injunction against the poisoning of wells as to make an immediate determination to oppose nuclear testing. No science can gauge the relative validity of these conflicting responses to the same facts.

Scientific method cannot determine whether the proponents of urban superhighways or those who complain about the resultant smog are in the right, or whether the benefits of nuclear tests to the national interest outweigh the hazards of fallout. No scientific principle can tell us how to make the choice, which may sometimes be forced upon us by the insecticide problem, between the shade of the elm tree and the song of the robin.

Certainly science can validly describe what is known about the information to be gained from a nuclear experiment, the economic value of a highway, or the hazard of radioactive contamination or of smog. The statement will usually be hedged with uncertainty, and the proper answer may sometimes be "We don't know," but in any case these separate questions do belong within the realm of science. However, the choice of the balance point between benefit and hazard is a value judgment; it is based on ideas of social good, on morality or religion—not on science.

In the "informed judgment" of which the Federal Radiation Council so properly speaks, the scientist can justly claim to be "informed," but he can make no valid claim for a special competence in "judgment." Once the scientific evidence has been stated, or its absence made clear, the establishment of a level of tolerance for a modern pollutant is a *social* problem and must be resolved by social processes. Thus the logic of the scientific problems which are raised by environmental pollution forces the resolution of these issues into the arena of public policy.

If resolutions of the problems created by the recent failures in large-scale technology require social judgments, who is to make these judgments? Obviously scientists must be involved in some way, if only because they have in a sense created the problems. But if these issues require social, political, and moral judgments, then they must also somehow reflect the demands, opinions, and ethics of citizens generally. Because new experiments and technological processes are so costly that the government must often pay for them, and because government officials mediate many social decisions, the government and its administrators are also involved. What are the proper roles of scientist, citizen, and administrator?

The Scientist's Role: Two Approaches

Since World War II scientists have become deeply concerned with public affairs. We are all acutely aware that our work, our ideas, and our daily activities impinge with a frightening immediacy on national politics, on international conflicts, on the planet's fate as a human habitation.

Scientists have tried to live with these responsibilities in a number of ways. Sometimes, in moments of impending crisis, we are aware only that the main outcome of science is that the planet has become a kind of colossal, lightly triggered time bomb. Then all we can do is to issue an anguished cry of warning. In calmer times we try to grapple with the seemingly endless problems of unraveling the tangle of nuclear physics, seismology, electronics, radiation biology, ecology, sociology, normal and pathological psychology, which, added to the crosscurrents of local, national, and international politics, has become the frightful chaos that goes under the disarming euphemism "public affairs."

Many scientists have studied the technology of the new issues and have mastered their vocabulary: megatonnage, micromicrocuries, threshold dose, and all the rest of the new technical terms. Nuclear physicists have struggled to learn the structure of the chromosome and how cows give milk. Biologists have returned to long-discarded text books of freshman physics.

A good deal of the scientist's concern for public issues may be generated by a sense of responsibility for the events which have converted nuclear energy from a laboratory experiment into the force which has almost alone molded the course of human events since 1945. It was a group of scientists who, fearful of the consequences of the possible development of nuclear weapons by the Nazis, conducted a strenuous campaign to convince the American government that they should be achieved in the United States first. As it turned out, Germany never succeeded in achieving an atomic bomb, and the Allies won the war against Germany without using it. Many of the scientists who worked on the United States atomic bomb were relieved to know that the threat which motivated them was gone and that the new force need never be used for destruction. But over their objections the weapon was turned against Japan, an enemy known to lack atomic arms. The human use of nuclear energy began with two explosions which took several hundred thousand lives; from this violent birth it has since grown into a destructive force of suicidal dimensions.

I believe that it is largely the weight of this burden which has caused the scientific community, since the end of the war, to examine with great care the interactions between science and society, to define the scientist's responsibilities to society, and to seek useful ways to discharge them.

For some time there has been a division of opinion within the scientific community on what responsibility the scientist has toward the social uses of his work. Some scientists have been guided by the idea that it is the scientist's duty to pursue knowledge of nature for its own sake without regard to social consequences. They believe that scientists, as scientists, have no special responsibility to foster any particular solution of the social issues that may result from their discoveries. They cling to the objectivity of the laboratory and try to keep their political views separate from their scientific duties. To other scientists such rigorous objectivity seems to imply a disregard for the nation's defense, or for the enormous destructiveness of nuclear war, or for the numerous ways in which science can serve human welfare. They seek to play a part in directing the power that they help to create.[2]

The second of these positions is relatively new and originates in scientists' intense concern with such dangerous issues as nuclear war. The rationale of this position appears to be approximately the following: Scientists have a particular moral responsibility to counter the evil consequences of their works. They are also in possession of the relevant technical facts essential to an understanding of the major public issues which trouble the world. Since scientists are trained to analyze the complex forces at work in such issues, they have an ability for rational thought which renders them to some degree detached from the emotions that encumber the ordinary citizen's views of these calamitous issues. What is more, some assert, the scientist is now in a particularly favorable position to be heard—by government executives, Congressional committees, the press, and the people at large—and therefore has important opportunities to influence social decisions.

This position, it will be noted, is quite neutral politically. It can justify strong statements for or against disarmament, civil defense, nuclear testing, or the space program. Such arguments in support of the scientist's inherent claim to the right of political leadership are the implicit background of a number of practical activities by scientists. These include publication of petitions and newspaper advertisements, political lobbying, and the operation of scientific auxiliaries to the major political parties.

This general approach has become fairly common in the scientific community, in contrast to the viewpoint, more predominant in the past, that a scientist's political role should be exercised apart from his professional one. One result is a growing tendency for considerations of scientific issues to appear with a strong admixture of political views. Witness the following examples from recent discussions, by scientists, of space research:

"On solid scientific grounds, on the basis of popular appeal, and in the interests of our prestige as a peace-loving nation capable of great scientific enterprise, exobiology's goals of finding and exploring extra-territorial life should be acclimed as the top-priority scientific goal of our space program."[3]

At the 1963 Senate hearings referred to in Chapter 4, a witness who has been a leader in developing the nation's scientific space program said, following an exposition of the scientific values of space research:

"Our goals in space provide to our Nation that spirit and momentum that avoids our collapse into the easy-going days that tolerate social abuse."[4]

In the same vein, I have seen a statement in which a number of distinguished scientists argued for a particular space experiment partly by supporting its scientific value, and partly by describing its special usefulness as a vehicle of international propaganda for the United States.

What is the harm in this approach? Why shouldn't scientists make effective use of their newly won position in society? Why not exploit political interests to further a scientific goal?

One danger has already been discussed—that the capability of science to understand nature, and to guide our efforts to control it, is being damaged by the pressure of political goals. And it is this capability alone that stands between us and man-made catastrophe. Clearly, no matter what else they do, scientists dare not act in such a way as to compromise the integrity of science or to damage its capability to seek the truth. But the notion that scientists have some special aptitude for the judgment of social issues—even of those which are due to the progress of science—runs a grave risk of damaging the integrity of science and public confidence in it. If two protagonists claim to know *as scientists*, through the merits of the methods of science, the one that nuclear testing is essential to the national interest, the other that it is destructive of the national interest, where lies the truth? I know from repetitious experience that the one question about fallout and nuclear war, and about the pesticide controversy, which most troubles the thoughtful citizen is: "How do I know which scientist is telling the truth?"

This is a painfully revealing question. It tells us that the public is no longer certain that scientists—all of them—"tell the truth," for otherwise why the question? The citizen has begun to doubt what he used to take for granted—that science is closely connected to the truth.

Now it seems to me that the citizen's confidence in the objectivity of science cannot be destroyed without disastrous consequences. We cannot really expect citizens, in general, to be capable of performing an independent check on the accuracy and validity of all of a scientist's statements about scientific matters. I am fully convinced that the citizen can and must study and come to understand the underlying facts about modern technological problems. But the citizen cannot check the calculations of the path of a rocket to the moon or

question the validity of the law of radioactive
decay. These conclusions he can accept, but only if
he knows that they *are* subject to the scrutiny of
scientists who will finally accept, reject, or modify
them. The citizen must take a good deal of science
as established by the simple fact that scientists
agree about it. It is therefore inevitable that unre-
solvable disagreement among scientists will erode
—and rightly so—the citizen's confidence in the
ability of science to get at the truth.

When scientists voice their social judgments
with the same authority that attaches to their professional pronouncements, the citizen is bound to
confuse the inevitable and insolvable disagreements with scientific disputes. If scientists attach
to their scientific conclusions those political views
or social judgments which happen to provide support for these conclusions, scientific objectivity
inevitably comes under a cloud.

In my opinion the notion that, because the
world is dominated by science, scientists have a
special competence in public affairs is also profoundly destructive of the democratic process. If
we are guided by this view, science will not only
create issues but also shield them from the cus-
tomary processes of administrative decision-
making and public judgment.

Nearly every facet of modern life is now so
encumbered by a tangled array of nuclear physics,
electronics, higher mathematics, and advanced
biology as to interpose an apparently insuperable
barrier between the citizen, the legislator, the administrator, and the major public issues of the
day. No one seems to be wholly exempt from this
estrangement. When, during the Congressional
hearings on the nuclear-test-ban treaty, Senator

Kuchel was confronted with a bewilderingly technical argument, he said in desperation to a scientist member of the AEC, "Let me put my tattered senatorial toga over your shoulders for a moment." When President Kennedy was questioned regarding government policy on the Starfish nuclear explosion, he was forced to fall back on the opinion of a scientist and closed the discussion with the remark, "After all, it's Dr. Van Allen's belt." Confronted by such examples, a citizen is likely to conclude that he must have a Ph.D. to support his judgments about nuclear war or the pesticide problem, or else be governed by the judgment of those who do.

The impact of modern science on public affairs has generated a nearly paralyzing paradox. Despite their origin in scientific knowledge and technological achievements (and failures) the issues created by the advance of science can only be resolved by moral judgment and political choice. But those who in a democratic society have the duty to make these decisions—legislators, government officials, and citizens generally—are often unable to perceive the issues behind the enveloping cloud of science and technology. And if those who have the special knowledge to comprehend the issues—the scientists—arrogate to themselves a major voice in the decision, they are likely to aggravate the very threats to the integrity of science which have helped to generate the problems in the first place.

A New Approach: An Informed Citizenry

There is no single magic word that solves this puzzle. But the same interlocking factors, looked at a little differently, do offer a solution: the scientist does have an urgent duty to *help* society solve the grave problems that have been created by the progress of science. But the problems are social and must be solved by social processes. In these processes the scientist has one vote and no claim to leadership beyond that given to any person who has the gift of moving his fellow men. But the citizen and the government official whose task it is to make the judgments cannot do so in the absence of the necessary facts and relevant evaluations. Where these are matters of science, the scientist as the custodian of this knowledge has a profound duty to impart as much of it as he can to his fellow citizens. But in doing so, he must guard against false pretensions and avoid claiming for science that which belongs to the conscience. By this means scientists can place the decisions on the grave issues which they have helped to create in the proper hands—the hands of an informed citizenry.

This is the view of the scientist's social responsibilities which was first developed by the AAAS Committee on Science in the Promotion of Human Welfare after a painstaking study of the conflict surrounding science and public policy. In its first report this committee said:

In sum, we conclude that the scientific community should on its own initiative assume an obligation to call to public attention those issues of public policy which relate to science, and to provide for the general public the facts and estimates of the effects of alternative policies which the citizen must have if he is to participate intelligently in the solution of these problems. A citizenry thus informed is, we believe, the chief assurance that science will be devoted to the promotion of human welfare.[5]

This view is plausible in its logic and laudable in its democratic intent. But does it work? An "informed citizenry" is the long-standing, and rarely attained, goal of social reformers. Most citizens, it can be argued, are unlikely to surmount the formidable difficulties involved in learning about fallout, pesticides, and air pollution. A scientist, once give a platform, might be unable to restrict himself to "the facts and estimates of the effects of alternative policies" and to refrain from also exhorting his audience to whatever policy he has adopted as his own. A scientist who has strong political sensibilities—and many do—may be unable to speak objectively on the data about nuclear war without developing a badly split personality. And by what means can scientists, who have no command over either the public news media or the machinery of government, overcome the governmental self-justification and journalistic inertia which so often impede public knowledge about complex, confused public affairs?

The Committee for Nuclear Information

Despite all these difficulties a growing number of scientists believe in this approach, because, like the preacher who believed in baptism, they have seen it done. Perhaps the most striking example is the work of the St. Louis Committee for Nuclear Information (CNI), which since 1958 has pioneered in public education about the science-generated issues of public affairs.[6]

Like many other communities in the United States and elsewhere in the world, St. Louis had been troubled and confused by fallout. Nuclear radiation, once a horrifying but (to everyone outside of Hiroshima and Nagasaki) distant prospect had come as close as the morning milk bottle. In 1957 St. Louis became one of six United States cities in which milk supplies were tested for strontium-90 by the Public Health Service. Its citizens learned, at intervals, that there was some number of "micromicrocuries" of strontium-90 in their milk, and that the amount was rising all over the country, with St. Louis often well in the lead.

St. Louis was besieged by numbers. In April 1958 St. Louis milk contained 6.5 micromicrocuries of strontium-90 per liter, but by July the number had increased to 18.7. Health authorities claimed that this amount was "insignificant" because the "maximum permissible concentration" (MPC), a level below which no medical effects were expected, was 80 micromicrocuries per liter. Then testimony at Congressional hearings revealed that, according to the International Commission on Radiological Protection, the MPC for large populations ought to be 30 micromicrocuries per liter and that levels would inevitably rise to that point if nuclear testing continued; in Mandan, North Dakota, milk had already reached the level of 33 micromicrocuries of strontium-90 per liter. Some experts contended that there was no absolutely safe level of radiation and that, according to accepted theories of radiation damage, any increase in radiation caused a proportional increase in the risk of medical damage, which might appear as leukemia or bone cancer.

Along with the confusion came concern. Mothers of young children besieged their pediatricians with questions: Was there really any danger? Should they give their children less milk? What about the strontium-90 content of other foods? Who was right about the MPC? Here was a clear need for an "informed citizenry."

Several of us in the St. Lous scientific community decided to try out the idea that scientists could usefully inform the public about such matters. In April 1958 an organization of several hundred St. Louisans was quietly formed. Citizens helped to define the areas of public concern. They also raised money, manned an office, and did the paper work. Physicians studied the medical problems of fallout. Physicists worked out the amounts and kinds of fallout expected from announced nuclear tests. Chemists concerned themselves with the distribution of the artificial radioisotopes of fallout in the environment. Biologists traced the passage of strontium-90 through the food chain and calculated the expected tissue damage. Seminars were held to exchange information; when the scientists had educated themselves they announced that they were available as public speakers on the facts about fallout.

The response was instantaneous—and overwhelming. When it became known that scientists were willing to try to explain the fallout problem, we were besieged with questions. We had to assure citizens that the white spots on their lawn grass were mold, not fallout; that there was no conceivable way to save the world by extending the half-life of radioactive atoms. And every lecture led to demands for more. Many of us became heavily engaged in the community's lecture circuit: PTAs, Lions, Rotarians, forums, television interviews. We discovered religious denominations and parts of our city that we never before knew existed. We saw at first hand the great need for providing the public with the information that every citizen must have if he is to decide for himself what policies our nation and the world should follow with respect to nuclear testing, civil defense, and disarmament.

At first, most of us tended to regard the task of public education as something apart from our professional life, as a kind of civic responsibility for which our general training prepared us. But, as we began to examine the available evidence for ourselves we found that it did not entirely support certain statements by government officials regarding, for example, the fallout hazard. And an alternative interpretation, presented to a lay audience, inevitably called forth the questions: "Why shouldn't the public be guided by the views of the government? Aren't the government scientists better informed than anyone else? Shouldn't we simply listen to what they say?"

CNI scientists found that the basic precept of scientific discourse—that data, interpretations, and conclusions be openly published and criticized—provided an effective answer to this difficulty. The story of the iodine-131 exposures near the Nevada nuclear test site is an example. Despite repeated government assurance of the "safety" of persons living near the test site, CNI scientists were not convinced. They obtained data from AEC and USPHS sources and calculated that the probable radiation dose to children in the area during certain tests in 1953 and later might have been as high as one thousand times the MPC. When this information was offered in testimony before the Joint Congressional Commission on Atomic Energy, the AEC responded with an outright and specific rejection of the CNI conclusions, even though an AEC scientist, H. A. Knapp, had just produced an official report which reached the same conclusion.[1]

This specific critique made it possible for CNI scientists to propose specific replies. For example, in defense of the failure to monitor directly for iodine-131 in the region of the test site, the AEC asserted that in the 1952-57 period there were no techniques available for such analyses. In rebuttal CNI scientists cited scientific publications as early as 1948 which described such techniques, and showed that one had actually been used by the AEC in 1953 to study sheep thyroid glands. In another instance, the AEC asserted that iodine-131 might not fall on pastures which are used by dairy cows, so that such fallout would not reach milk and therefore not harm children. In reply CNI produced a table reporting milk-cow census data for the Utah region, which showed that there were 6,000 dairy cows distributed among several hundred farms in the area. Along with a series of detailed criticisms, CNI sent the AEC Division of Operational Safety a letter asserting that "either you or we are wrong" and asking for evidence of CNI's error or an acknowledgment that the group's analysis of the iodine-131 situation in Utah was correct. Although no reply was forthcoming from the AEC, the net impact of the discussion on other interested parties was noteworthy: The U.S. Public Health Service agreed with the CNI proposal for a medical survey of the children in the Utah area, and such a survey began, with the results described in Chapter 1.

Often the data relevant to science-generated public issues are in fact made fully public by government agencies, but in such technical terms as to be meaningless to the general public. In 1959 the Joint Congressional Committee on Atomic Energy held a series of detailed hearings on the consequence of a nuclear war, assuming a plausible size and distribution of the attack. Witness after witness gave details about the death and destruction of a nuclear holocaust. Although the hearings reported the first comprehensive picture of the potential catastrophe which has so strongly shaped our political lives, most newspaper reports gave only brief mention to a few highlights. The information was there, but the public was not well informed.

CNI arranged with the Joint Committee and with some of the witnesses to obtain copies of the testimony as it was delivered. A team of CNI scientists studied the testimony and prepared summaries of the physical, ecological, and medical effects of the two nuclear bombs that the Joint Committee's assumed nuclear attack had assigned to St. Louis. The members of the group recognized that scientific data on a subject so remote—and so alarming—as a hypothetical nuclear bombing of a city might be difficult to cast into a readable form but believed that the information was nevertheless of commanding importance. After several unsatisfactory drafts had been prepared they decided that the scientific reports could be translated without loss of detail or accuracy into fictional accounts by supposed survivors of the hypothetical bombing of St. Louis, documented with footnotes relating the facts described to the Congressional hearings. The result was the publication in the CNI monthly magazine *Nuclear Information* (now *Scientist and Citizen*) of a distinctive, moving, yet accurate account of what the evidence presented at the Congressional hearings had to say about the effects of a nuclear attack on a city.[9] In addition to the regular circulation of its publication, CNI distributed, on request, some 45,000 copies of this single issue. It was reprinted in the *Saturday Review*, the *St. Louis Post-Dispatch*, the *Houston Post*, the *San Francisco News-Call Bulletin*, and in a dozen other periodicals in the United States, Great Britain, Canada, Mexico, France, Sweden, and Denmark.

Here the CNI scientists had served as translators—bridging the morass of technical terms and scientific abstractions which separates most citizens from the awful fact of nuclear war. And apparently even trained government officials needed CNI's translation service to appreciate what went on at the Congressional hearings: On request CNI distributed 50 copies of the article to the Red Cross, 800 to various branches of the military services, 1,000 to the Food and Drug Administration, and 2,000 to the Department of Health, Education, and Welfare. About 2,500 copies were ordered by state and federal civil defense organizations. The War College and the Air University of the United States Air Force received permission to reprint the bulletin for the use of their graduating classes.

Does such information do any good? Even if informed, what can the citizen do about established government policy? A good test of this question is the story of Project Chariot, a proposal put forward by the AEC in 1958 to blast out a new harbor on the northwest shore of Alaska with hydrogen bombs. Soon after the proposal was announced, a controversy arose regarding radiation effects on local flora and fauna and on the life of the nearby Eskimo villages. CNI asked the AEC for information. The AEC replied that while no reports on the biological hazards were available, a number of biologists, whose names were provided, were engaged in such studies under AEC sponsorship. CNI wrote to these biologists for information. Several at the University of Alaska replied with copies of reports that they had earlier submitted to the AEC. Study of these soon showed that the Alaskan investigations had uncovered a series of important biological difficulties in Project Chariot.

For example, the AEC had apparently selected the site of the proposed explosion because it was in the center of a long reach of uninhabited shoreline. But what had not been taken into account was that inhabitants from two Eskimo villages, although distant from the site, intensively hunted and gathered food in the "empty" region; vast food-gathering areas are essential for survival in arctic areas.

CNI invited the Alaskan biologists to prepare articles for publication in *Nuclear Information* [*Scientist and Citizen*]. To these articles were added studies by CNI physicists of some of the problems involved in such earth-moving projects. The result was an extensive analysis of Project Chariot, which showed, among other things, that it would seriously endanger the food supply of the local Eskimo settlements. At about the same time the AEC issued a report on the state of the project, which gave scant attention to these biological difficulites. For several years Alaskans had been assured by AEC representatives that there was no danger from the explosions. Now it appeared that there was, on scientific grounds, another side to the story.

But could such information really be brought to bear on a governmental apparatus that had already committed millions of dollars to the project? The people most concerned were the inhabitants of the several isolated, very primitive Eskimo settlements along the west coast of Alaska. Few of the villagers could read or speak English. They even lacked a written language of their own. But with the nuclear age we also have the tape recorder, with consequences that have been reported by Dr. W. H. Pruitt, one of the Alaskan biologists who studied the ecological effects of the Chariot proposal:

> The tape recorder is one of the greatest inventions as far as the Eskimo culture is concerned. Every village has several tape recorders. . . . There is a constant traffic in tapes from one village to another. . . . The Chariot information and the material contained in the Alaska Conservation Society and CNI bulletins got into the tapes and it literally swept the Arctic coast, from Kakhtovik all the way down to Nome and below. . . . I recall, also, meeting an Eskimo driving a dog team on the trail one time, and, by golly, he had a copy of the CNI bulletin tucked inside his parka.[11]

It is risky to claim a particular cause for any given event in political life, but the record does show that the Chariot proposal met little opposition until independent scientists of the Alaska Conservation Society and CNI began to explain some of the biological consequences—and that the project was finally killed.

Public information developed by independent scientists has figured in a number of controversies over nuclear projects. When the State of Massachusetts proposed to establish a nuclear-waste-processing plant on Cape Cod, scientists at the Marine Biological Laboratory at Woods Hole assembled information showing that special hazards would result from the geological situation peculiar to the region. After public hearings at which these facts were aired, the project was abandoned. The proposal to build a large nuclear reactor at Bodega Bay in California ran into stiff opposition based largely on evidence that the reactor would be so close to the San Andreas earthquake fault as to raise a serious hazard of an earthquake-induced radiation catastrophe. Many people were surprised at the ease with which the 1963 test-ban treaty was approved by the United States Senate. Several observers have noted a possible connection with the numerous letters received by Senators from housewives and mothers who not only wanted the treaty approved but could cite serious scientific grounds for their belief that it would help reduce the medical hazards from fallout.

In a surprising way CNI's program has turned out to be a two-way street. While citizens are usually on the receiving end of information provided by scientists, in at least one important instance in connection with fallout citizens have themselves produced new scientific information. In 1958 the prominent American biochemist Herman Kalckar published a paper in the international scientific journal *Nature* on the problem of assessing the strontium-90 uptake in children's bones. The importance of such information was widely recognized, since there was no basis from previous experience for an accurate prediction of how much of the strontium-90 in the diet would remain in the bones—where it might, if sufficiently concentrated, in time induce leukemia or bone tumors. Methods were available for strontium-90 analysis of bones, but since the analysis could only be made on corpses the data were scattered and incomplete. Dr. Kalckar pointed out that strontium-90 is also deposited in the teeth, that the milk teeth are shed after a time and could easily be analyzed for strontium-90, and that the results could be directly correlated with the situation in bone.

The idea was sound but difficult to carry out. A single tooth contains so little strontium-90 that considerable numbers of teeth would be needed to produce statistically significant results. The teeth would need to be collected regularly over a long period in order to make up for the delay of about 5 to 7 years between the time a milk tooth is formed and the time it is shed. When CNI scientists read Dr. Kalckar's paper, they estimated that the project would require the collection of about 50,000 teeth a year for at least ten years to produce useful results.

But CNI members were struck with the idea and enthusiastic about "doing something" about fallout, if only providing scientific information about it. In December 1958 CNI organized the Baby Tooth Survey[12] and with the help of numerous willing mothers and children began collecting teeth. When the project began no one was sure how so many teeth could be collected, nor did CNI have the resources to carry out the elaborate and expensive analyses for strontium-90. Within a year the donation of baby teeth had become a way of life among St. Louis children. By 1966 the Baby Tooth Survey had collected over 200,000 teeth. With CNI's cooperation the School of Dentistry at Washington University established an analytical laboratory, supported by a grant from the U.S. Public Health Services. The laboratory has published the first, and thus far the only detailed studies of the absorption of strontium-90 by a large population of children through the major period of nuclear testing. It has provided scientists with unique data on this new problem. The children who gave up the traditional visit of the tooth fairy to contribute their baby teeth to science were in this case helping to develop new scientific information for their own future welfare.

Much has been written about the alienation of citizens from the complexities of modern public affairs. Experts have everywhere intruded between the issues and the public. The Jeffersonian concept of an educated, informed electorate appears to be a naïve and distant ideal. But at least in one area—science and technology, on which so much of our future depends—an effort is being made to make the ideal a reality.

Margaret Mead, the anthropologist, has called scientists' efforts to alert citizens to these issues and provide them with information needed for evaluating the benefits and hazards of modern technology "a new social invention." This may turn out to be the one invention of our technological age which can conserve the environment and preserve life on the earth.

7

To Survive on the Earth

ONCE the problems are perceived by science, and scientists help citizens to understand the possible solutions, what actions can be taken to avoid the calamities that seem to follow so closely on the heels of modern technological progress? I have tried to show that science offers no "objective" answer to this question. There is a price attached to every solution; any judgment will necessarily reflect the value we place on the benefits yielded by a given technological advance and the harm we associate with its hazards. The benefits and the hazards can be described by scientific means, but each of us must choose that balance between them which best accords with our own belief of what is good—for ourselves, for society, and for humanity as a whole.

In discussing what ought to be done about these problems, I can speak only for myself. As a scientist, I can arrive at my own judgments—subject to the open criticism which is so essential to scientific discourse—about the scientific and technological issues. As a citizen, I can decide which of the alternative solutions my government ought to pursue, and, using the instruments of politics, act for the adoption of this course. As a human being, I can express in this action my own moral convictions.

As a biologist, I have reached this conclusion: we have come to a turning point in the human habitation of the earth. The environment is a complex, subtly balanced system, and it is this integrated whole which receives the impact of all the separate insults inflicted by pollutants. Never before in the history of this planet has its thin life-supporting surface been subjected to such diverse, novel, and potent agents. I believe that the cumulative effects of these pollutants, their interactions and amplification, can be fatal to the complex fabric of the biosphere. And, because man is, after all, a dependent part of this system. I believe that continued pollution of the earth, if unchecked, will eventually destroy the fitness of this planet as a place for human life.

My judgment of the possible effects of the most extreme assault on the biosphere—nuclear war—has already been expressed. Nuclear war would, I believe, inevitably destroy the economic, social, and political structure of the combatant nations; it would reduce their populations, industry and agriculture to chaotic remnants, incapable of supporting an organized effort for recovery. I believe that world-wide radio-active contamination, epidemics, ecological disasters, and possibly climatic changes would so gravely affect the stability of the biosphere as to threaten human survival everywhere on the earth.

If we are to survive, we need to become aware of the damaging effects of technological innovations, determine their economic and social costs, balance these against the expected benefits, make the facts broadly available to the public, and take the action needed to achieve an acceptable balance of benefits and hazards. Obviously, all this should be done *before* we become massively committed to a new technology. One of our most urgent needs is to establish within the scientific community some means of estimating and reporting on the expected benefits and hazards of proposed environmental interventions *in advance*. Such advance consideration could have averted many of our present difficulties with detergents, insecticides, and radioactive contaminants. It could have warned us of the tragic futility of attempting to defend the nation's security by a means that can only lead to the nation's destruction.

We have not yet learned this lesson. Despite our earlier experience with nondegradable detergents, the degradable detergents which replaced them were massively marketed, by joint action of the industry in 1965, without any pilot study of their ecological effects. The phosphates which even the new detergents introduce into surface waters may force their eventual withdrawal. The United States, Great Britain, and France are already committed to costly programs for supersonic transport planes but have thus far failed to produce a comprehensive evaluation of the hazards from sonic boom, from cosmic radioactivity, and from the physiological effects of rapid transport from one time zone to another. The security of every nation in the world remains tied to nuclear armaments, and we continue to evade an open public discussion of the basic question: do we wish to commit the security of nations to a military system which is likely to destroy them?

It is urgent that we face this issue openly, now, before by accident or design we are overtaken by nuclear catastrophe. U Thant has proposed that the United Nations prepare a report on the effects of nuclear war and disseminate it throughout the world. Such a report could become the cornerstone of world peace. For the world would then know that, so long as nuclear war remains possible, we are all counters in a colossal gamble with the survival of civilization.

The costs of correcting past mistakes and preventing the threatened ones are already staggering, for the technologies which have produced them are now deeply embedded in our economic, social, and political structure. From what is now known about the smog problem, I think it unlikely that gasoline-driven automobiles can long continue to serve as the chief vehicle of urban and surburban transportation without imposing a health hazard which most of us would be unwilling to accept. Some improvement will probably result from the use of new devices to reduce emission of waste gasoline. But in view of the increasing demand for urban transportation any really effective effort to reduce the emission of waste fuel, carbon monoxide, and lead will probably require electric-powered vehicles and the replacement of urban highway systems by rapid transit lines. Added to current demands for highway-safe cars, the demand for smog-free transportation is certain to have an important impact on the powerful and deeply entrenched automobile industry.

The rapidly accelerating pollution of our surface waters with excessive phosphate and nitrate from sewage and detergents will, I believe, necessitate a drastic revision of urban waste systems. It may be possible to remove phosphates effectively by major modifications of sewage and water treatment plants, but there are no methods in sight that might counter the accumulation of nitrate. Hence, control will probably need to be based chiefly on preventing the entry of these pollutants into surface waters.

According to a report by the Committee on Pollution, National Academy of Sciences,[3] we need to plan for a complete transformation of urban waste-removal systems, in particular to end the present practice of using water to get rid of solid wastes. The technological problems involved are so complex that the report recommends, as an initial step, the construction of a small pilot city to try out the new approach.

The high productivity of American agriculture, and therefore its economic structure, is based on the use of large amounts of mineral fertilizer in which phosphate and nitrate are major components. This fertilizer is not entirely absorbed by the crops and the remainder runs off into streams and lakes. As a result, by nourishing our crops and raising agricultural production, we help to kill off our lakes and rivers. Since there is no foreseeable means of removing fertilizer runoff from surface waters, it will become necessary, it seems to me, to impose severe restrictions on the present unlimited use of mineral fertilizers in agriculture. Proposed restraints on the use of synthetic pesticides have already aroused a great deal of opposition from the chemical industry and from agriculture. Judged by this response, an attempt to regulate the use of mineral fertilizers will confront us with an explosive economic and political problem.

And suppose that, as it may, the accumulation of carbon dioxide begins to threaten the entire globe with catastrophic floods. Control of this danger would require the modification, throughout the world, of domestic furnaces and industrial combustion plants—for example, by the addition of devices to absorb carbon dioxide from flue gases. Combustion-driven power plants could perhaps be replaced with nuclear ones, but this would pose the problem of safely disposing of massive amounts of radioactive wastes and create the hazard of reactor accidents near centers of population. Solar power, and other techniques for the production of electrical power which do not require either combustion or nuclear reactors, may be the best solution. But here too massive technological changes will be needed in all industrial nations.

The problems of industrial and agricultural pollution, while exceedingly large, complex, and costly, are nevertheless capable of correction by the proper technological means. We are still in a period of grace, and if we are willing to pay the price, as large as it is, there is yet time to restore and prserve the biological quality of the environment. But the most immediate threat to survival— nuclear war—would be a blunder from which there would be no return. I know of no technological means, no form of civil defense or counteroffensive warfare, which could reliably protect the biosphere from the catastrophic effects of a major nuclear war. There is, in my opinion, only one way to survive the threats of nuclear war—and that is to insure that it never happens. And because of the appreciable chance of an accidental nuclear way, I believe that the only way to do so is to destroy the world's stock of nuclear weapons and to develop less self-defeating means of protecting national security. Needless to say, the political difficulties involved in international nuclear disarmament are monumental.

Despite the dazzling successes of modern technology and the unprecedented power of modern military systems, they suffer from a common and catastrophic fault. While providing us with a bountiful supply of food, with great industrial plants, with high-speed transportation, and with military weapons of unprecedented power, they threaten our very survival. Technology has not only built the magnificent material base of modern society, but also confronts us with threats to survival which cannot be corrected unless we solve very grave economic, social, and political problems.

How can we explain this paradox? The answer is, I believe, that our technological society has committed a blunder familiar to us from the nineteenth century, when the dominant industries of the day, especially lumbering and mining, were successfully developed—by plundering the earth's natural resources. These industries provided cheap materials for constructing a new industrial society, but they accumulated a huge debt in destroyed and depleted resources, which had to be paid by later generations. The conservation movement was created in the United States to control these greedy assaults on our resources. The same thing is happening today, but now we are stealing from future generations not just their lumber or their coal, but the basic necessities of life: air, water, and soil. A new conservation movement is needed to preserve life itself.

The earlier ravages of our resources made very visible marks, but the new attacks are largely hidden. Thoughtless lumbering practices left vast scars on the land, but thoughtless development of modern industrial, agricultural, and military methods only gradually poison the air and water. Many of the pollutants—carbon dioxide, radioisotopes, pesticides, and excess nitrate—are invisible and go largely unnoticed until a lake dies, a river becomes foul, or children sicken. This time the world is being plundered in secret.

The earlier depredations on our resources were usually made with a fair knowlege of the harmful consequences, for it is difficult to escape the fact that erosion quickly follows the deforestation of a hillside. The difficulty lay not in scientific ignorance, but in willful greed. In the present situation, the hazards of modern pollutants are generally not appreciated until after the technologies which produce them are well established in the economy. While this ignorance absolves us from the immorality of the knowingly destructive acts that characterized the nineteenth century raids on our resources, the present fault is more serious. It signifies that the capability of science to guide us in our interventions into nature has been seriously eroded—that science has, indeed, got out of hand.

In this situation, scientists bear a very grave responsibility, for they are the guardians of the integrity of science. In the last few decades serious weaknesses in this system of principles have begun to appear. Secrecy has hampered free discourse. Major scientific enterprises have been governed by narrow national aims. In some cases, especially in the exploration of space, scientists have become so closely tied to basically political aims as to relinquish their traditional devotion to open discussion of conflicting views on what are often doubtful scientific conclusions.

What can scientists do to restore the integrity of science and to provide the kind of careful technology that is essential if we are to avoid catastrophic mistakes? No new principles are needed; instead, scientists need to find new ways to protect science itself from the encroachment of political pressures. This is not a new problem, for science and scholarship have often been under assault when their freedom to seek and to discuss the truth becomes a threat to existing economic or political power. The internal strength of science and its capability to understand nature have been weakened whenever the priciples of scientific discourse were compromised, and restored when these principles were defended. The medieval suppressions of natural science, the perversion of science by Nazi racial theories, Soviet restraints on theories of genetics, and the suppression by United States military secrecy of open discussion of the Starfish project, have all been paid for in the most costly coin—knowledge. The lesson of all these experiences is the same. If science is to perform its duty to society, which is to guide, by objective knowledge, human interactions with the rest of nature, its integrity must be defended. Scientists must find ways to remove the restraints of secrecy, to insist on open discussion of the possible consequences of large-scale experiments *before* they are undertaken, to resist the hasty and unconditional support of conclusions that conform to the demands of current political or economic policy.

Apart from these duties toward science, I believe that scientists have a responsibility in relation to the technological uses which are made of scientific developments. In my opinion, the proper duty of the scientist to the social consequence of his work cannot be fulfilled by aloofness or by an approach which arrogates to scientists alone the social and moral judgments which are the right of every citizen. I propose that scientists are now bound by a new duty which adds to and extends their older responsibility for scholarship and teaching. We have the duty to inform, and to inform in keeping with the traditional principles of science, taking into account all relevant data and interpretations. This is an involuntary obligation to society; we have no right to withhold information from our fellow citizens, or to color its meaning with our own social judgments.

Scientists alone cannot accomplish these aims, for despite its tradition of independent scholarship, science is a dependent segment of society. In this sense defense of the integrity of science is a task for every citizen. And in this sense, too, the fate of science as a system of objective inquiry, and therefore its ability safely to guide the life of man on eath, will be determined by social intent. Both awareness of the grave social issues generated by new scientific knowledge, and the policy choices which these issues require, therefore become matters of public morality. Public morality will determine whether scientific inquiry remains free. Public morality will determine at what cost we shall enjoy freedom from insect pests, the convenience of automobiles, or the high productivity of agriculture. Only public morality can determine whether we ought to intrust our national security to the catastrophic potential of nuclear war.

There is a unique relationship between the scientist's social responsibilities and the general duties of citizenship. If the scientist, directly or by inferences from his actions, lays claim to a special responsibility for the resolution of the policy issues which relate to technology, he may, in effect, prevent others from performing their own political duties. If the scientist fails in his duty to inform citizens, they are precluded from the gravest acts of citizenship and lose their right of conscience.

We have been accustomed, in the past, especially in our organized systems of morality—religion—to exemplify the principles of moral life in terms which relate to Egypt under the pharaohs or Rome under the emperors. Since the establishment of Western religions, their custodians have, of course, labored to achieve a relevance to the changing states of society. In recent times the gap between traditional moral principles and the realities of modern life has become so large as to precipitate, beginning in the catholic church, and less spectacularly in other religious denominations, urgent demands for renewal—for the development of statements of moral purpose which are directly relevant to the modern world. But in the modern world the substance of moral issues cannot be perceived in terms of the casting of stones or the theft of a neightbor's ox. The moral issues of the modern world are embedded in the complex substance of science and technology. The exercise of morality now requires the determination of right between the farmers whose pesticides poison the water and the fishermen whose livelihood may thereby be destroyed. It calls for a judgment between the advantage of replacing a smoky urban power generator with a smoke-free nuclear one which carries with it some hazard of a catastrophic accident. The ethical principles involved are no different from those invoked in earlier times, but the moral issues cannot be discerned unless the new substance in which they are expressed is understood. And since the substance of science is still often poorly perceived by most citizens, the technical content of the issues of the modern world shields them from moral judgment.

Nowhere is this more evident than in the case of nuclear war. The horrible face of nuclear war can only be described in scientific terms. It can be pictured only in the language of roentgens megatonnage; it can be understood only by those who have some appreciation of industiral organization, of human biology, of the intricacies of world-wide ecology. The self-destructiveness of nuclear war lies hidden behind a mask of science and technology. It is this shield, I believe, which has protected this most fateful moral issue in the history of man from the judgment of human morality. The greatest moral crime of our time is the concealment of the nature of nuclear war, for it deprives humanity of the solemn right to sit in judgment on its own fate; it condemns us all, unwittingly, to the greatest dereliction of conscience.

The obligation which our technological society forces upon all of us, scientist and citizen alike, is to discover how humanity can survive the new power which science has given it. It is already clear that even our present difficulties demand far-reaching social and political actions. Solution of our pollution problems will drastically affect the economic structure of the automobile industry, the power industry, and agriculture and will require basic changes in urban organization. To remove the threat of nuclear catastrophe we will be forced at last to resolve the pervasive international conflicts that have bloodied nearly every generation with war.

Every major advance in the technological competence of man has enforced revolutionary changes in the economic and political structure of society. The present age of technology is no exception to this rule of history. We already know the enormous benefits it can bestow; we have begun to perceive its frightful threats. The political crisis generated by this knowledge is upon us.

Science can reveal the depth of this crisis, but only social action can resolve it. Science can now serve society by exposing the crisis of modern technology to the judgment of all mankind. Only this judgement can determine whether the knowledge that science has given us shall destroy humanity or advance the welfare of man.

Reference Notes

1: Is Science Getting Out of Hand?

1. For an excellent summary of the events associated with the great Northeast blackout and several observers' comments on how the blackout affected their view of the reliability of modern science and technology see *The Night the Lights Went Out* by the Staff of *The New York Times* (New York: New American Library, 1965). Although the "cause" of the blackout has been discovered in the sense that the failure began with the opening of an incorrectly set relay at Queenston, Ontario, the crucial point is that no one yet knows why that failure precipitated the spreading disaster. In other words, the power network was established before anyone understood the circumstances under which it would fail.

2. *Life*, November 19, 1965.

3. T. H. White, *Life, loc. cit.*

4. Norbert Wiener, "Some Moral and Technical Consequences of Automation," *Science*, 131:3410 (May 6, 1960), p. 1355.

5. For a detailed account of the iodine-131 story in the areas near the Nevada test site see *Scientist and Citizen*, August 1963. Recently Dr. Charles W. Mays, of the University of Utah, has indicated that iodine-131 levels comparable to those occurring in Washington County may have affected a much larger area of Utah.

6. *The Thirteenth Semiannual Report of the*

Atomic Energy Commission (Washington, D.C.: Government Printing Office, January 1953) , p. 124.

2: *Sorcerer's Apprentice*

1. President's Science Advisory Committee, *Restoring the Quality of Our Environment* (Washington, D.C.: Government Printing Office, November 1965) . This report is an excellent source of detailed information on environmental contamination.

2. Comprehensive information about contamination from fallout resulting from nuclear tests can be found in a number of issues of *Science and Citizen* (published by the St. Louis Committee for Nuclear Information) , especially September/October 1964 and September 1965.

3. The quotation is contained in an A.P. dispatch from Washington, D.C., dated October 24, 1956.

4. The quotation is from his televised address of October 12, 1964.

5. Readers interested in a more optimistic appraisal of fallout hazards will find it in *The Legacy of Hiroshima*, by Edward Teller and Allen Brown (New York: Doubleday, 1962) . For a critical review of this book, in which I have pointed out some of the scientific errors that led to the authors' optimism about fallout, see *Chemical and Engineering News*, 40:86 (1962) .

6. John M. Fowler, *Fallout* (New York: Basic Books, 1960) , p. 30.

7. See *Hearings, Special Subcommittee on Radiation*, Joint Congressional Committee on Atomic Energy, June 4-7, 1957, Part 2, p. 1489 for Dr. Libby's estimate. See Hearings of the same subcommittee, May 1959, for the 1958 measurements.

8. *The Thirteenth Semiannual Report of the Atomic Energy Commission, op. cit.*, p. 122.

9. *The Thirteenth Semiannual Report of the Atomic Energy Commission, op. cit.*, p. 78.

10. Because of the considerable literature already

available on the hazards of insecticides, this issue is treated extremely briefly here. A valuable summary of the ecological effect of pesticides is given in *Pesticides and the Living Landscape* by R. C. Rudd (Madison, Wis.: University of Wisconsin Press, 1964). And of course no one interested in this subject should fail to read *Silent Spring* by Rachel Carson (Boston: Houghton Mifflin, 1962).

3: Greater Than the Sum of its Parts

1. Typical of recent enthusiasm about achievements in molecular biology and related research fields is J. M. Barry, *Molecular Biology: Genes and the Chemical Control of Living Cells* (Englewood Cliffs, N.J.: Prentice-Hall, 1964).

2. The biological significance of DNA has been treated in a considerable amount of popular writing of late. Readers who are interested in a summary of the orthodox approach to this problem, that is, that DNA *is* the secret of life, might begin with the following: George Beadle and Muriel Beadle, *The Language of Life: On Introduction to the Science of Genetics* (Garden City, N.Y.: Doubleday, 1966); F. H. C. Crick, "The Structure of the Hereditary Material," *Scientific American*, October 1954, and "The Genetic Code," *Scientific American*, October 1962; and Franklin W. Stahl, *The Mechanics of Inheritance* (Englewood Cliffs, N.J.: Prentice-Hall, 1964).

3. John B. Graham, *Federation Proceedings*, 24 (1965), p. 1237.

4. *Genetics: Genetic Information and the Control of Protein Structure and Function*, Transactions of the First Conference, October 19-22, 1959 (Madison, N.J.: Madison Printing, 1959).

5. For an excellent and remarkably understandable account of the successful theoretical analysis of superconductivity see J. B. Bardeen, "Development of Concepts in Superconductivity, *Science Today*, January 1963, pp. 19-28.

6. I must record my debt to my good friend Martin Quigley, who suggested that the view developed here can be succinctly summarized in this aphorism.

7. George Gaylord Simpson, "The Biological Nature of Man," *Science*, April 22, 1966, pp. 472-478. For a more extensive discussion of Simpson's views on these matters, the reader should see his collection of essays entitled *This View of Life* (New York: Harcourt, Brace & World, 1963). For another explanation by a most distinguished biologist of the view that molecular biology does not explain the unique properties of life, see *The Dreams of Reason* by René Dubos (New York: Columbia University Press, 1961).

For a devastating critique of current oversimplifications in the field of molecular genetics, written by one of the pioneers in nucleic acid biochemistry, see Edwin Chargaff, *Essays on Nucleic Acids* (New York: Elsevier Press, 1963).

For a detailed presentation of the important evidence on the biological roles of DNA which is not encompassed by the orthodox theory and of evidence relevant to the alternative view of the role of DNA presented in this chapter see the following papers by the present author: "The Roles of Deoxyribonucleic Acid in Inheritance," *Nature*, June 6, 1964: "Deoxyribonucleic Acid and the Molecular Basis of Self-Duplication," *Nature*, August 1, 1964; "DNA and the Chemistry of Inheritance," *American Scientist*, September 1964.

8. Isaac Asimov, *The Intelligent Man's Guide to Modern Science*, Vol. 2 (New York: Basic Books, 1960).

4: Science versus Society

1. The material discussed in this chapter is closely related to a report of the Committee on Science in the Promotion of Human Welfare of the American Association for the Advancement of Science entitled "The Integrity of Science," published in *American*

Scientist, June 1965. This report was the result of a detailed study of recent instances of large-scale experimentation. Detailed evidence in support of most of the conclusions reached in this chapter will be found in that report. The earlier reports of that committee are also relevant to much of the material discussed in this book: "Science and Human Welfare," *Science*, July 8, 1960; "Science and Human Survival," *Science*, December 29, 1961; "Science and the Race Problem," *Science*, November 1, 1963.

2. The relative growth rate of scientific research in comparison to other segments of our social structure is illustrated by the following data: Between 1959 and 1964 the U.S. military budget increased about 14 percent, while government expenditures for basic research increased about threefold.

3. There are important differences between *basic* and *applied* science which need to be taken into account in any consideration of the effects of secrecy on scientific progress. What is under discussion here is solely the impact of secrecy on progress in *basic* research, where it is clear that secrecy has effects which are so harmful as to outweigh any possible temporary advantages to the military or to any other areas of applied science. Quite another matter is the problem of secrecy imposed on the *applications* of basic scientific knowledge to specific technological problems. Of course, here too progress in the general sense would certainly be enhanced by widespread dissemination of results. However, it has been proposed, for example in connection with the problems involved in the patent system, that either exclusive knowledge or the exclusive right to use knowledge in connection with technological developments may be necessary to provide an investment incentive. This is a complicated problem which cannot be resolved without a detailed discussion of the economic background of technological progress in different forms of society, but these issues are not under discussion here.

4. Quoted in a press release issued by the AEC and the Department of Defense, Washington, D.C., May 28, 1962.

5. C. E. McIlwain, "The Radiation Belts, Natural and Artificial," *Science*, October 18, 1963.

6. One of the curious anomalies that occurred when some fallout data were first declassified was the publication of papers in the open scientific literature which included references to government documents restricted by a secrecy classification. It is perhaps useful for the reader of a scientific paper to know that a secret document exists, but the frustration felt by those of us who needed to rely on open publication to comprehend the fallout problem is understandable.

7. No one should conclude that the contributions made by independent scientists to our understanding of the fallout problem were more important than the enormous amount of work done by government scientists, even during the periods of utmost secrecy about fallout. Quantitatively, the great bulk of fallout information was derived from many excellent scientific investigations carried out by government laboratories and by independent laboratories under contract to government agencies. The point being made here is that certain omissions and errors were inevitable in the secret development of such a complex field and that these were discovered, quite naturally, when the work of government scientists was exposed to the scrutiny of a much larger scientific community.

8. S. V. Dalgaard-Mikkelsen and E. Poulsen, *Pharmacological Reviews*, June 1962.

The problem of evaluating the effects of industrial secrecy on our understanding of environmental issues is of course a very considerable one and needs careful study before its full impact is appreciated. That industrial secrecy has had some effect and that this has certainly been in the direction of restricting broad knowledge of environmental hazards from such chemicals is, I believe, self-evident.

9. *Hearings before the Committee on Aeronautical and Space Sciences,* United States Senate, June 10 and 11, 1963 (Washington, D.C.: Government Printing Office). Each of the scientists who appeared before the committee as a witness was asked a series of specific questions, including several on the influence of space research on national prestige. Most of the witnesses strongly favored intensive support of the space program; a distinguished exception is the testimony of Dr. Polykarp Kusch, which appears on pages 67-73 of the Hearings Report.

10. The quotation cited here is from *A Review of Space Research* (Washington, D.C.: Publication 1079 of the National Research Council, National Academy of Sciences, 1962). For a summary of the results of the questionnaire from the American Association for the Advancement of Science which revealed the relatively large number of scientists supported by NASA funds see *Science,* July 24, 1964.

11. *Hearings before the Committee on Aeronautical and Space Sciences,* United States Senate, November 21 and 22, 1963 (Washington, D.C.: Government Printing Office), p. 34.

12. The priorities originally accorded to various space projects by the Space Science Board can be found in *Science in Space,* edited by Lloyd V. Berkner and Hugh Odishaw, respectively chairman and executive director of the Space Science Board, National Academy of Sciences (New York: McGraw-Hill, 1961). Although the book was published in 1961, the priorities accorded various projects refer to Space Science Board decisions made in 1959, before President Kennedy established the Apollo project as the highest-priority activity of the space program.

13. Address by Homer E. Newell, Director, Office of Space Sciences, NASA, at the annual meeting of the American Association for the Advancement of Science at Philadelphia on December 26, 1962.

14. "The Integrity of Science" (*loc. cit.*). The quotations which follow are from this report.

5: The Ultimate Blunder

1. The quotations by President Kennedy and former Premier Khrushchev may be found in their addresses before the United Nations General Assembly, September 18 and 25, 1961, respectively.

2. Pope John XXIII, *Pacem in Terris* (National Catholic Welfare Conference, 1963), p. 27.

3. Steuart L. Pittman, "Civil Defense in a Balanced National Security," *Bulletin of the Atomic Scientists,* June 1964.

4. For a detailed summary of the facts relative to the effects of nuclear war see a series of reports published in the following issues of *Scientist and Citizen:* July 1962; October-November 1962; December 1962; February 1963; May 1963; July 1963; September-October 1963; February 1964; March 1964; June-July 1964; May-June 1965; August 1965; February-March 1966.

5. The relative blast and fire effects of an exploding nuclear bomb depend a great deal on the altitude at which it explodes, the weather conditions, and the nature of the target area. The facts cited here are purely illustrative and refer to explosions at a height chosen to achieve maximum destruction.

6. The rate at which fallout particles return to earth is affected in complex ways by a number of factors, including the geographical location of the explosion, local weather conditions at the time, the time of year, and large-scale weather patterns.

7. The figures cited in connection with the destructive effects of a stated number of nuclear bombs on various types of targets are based on the assumption that each bomb strikes at the center of the target area. Less than perfect accuracy would of course reduce the destructive effects. However, this is probably not a very large influence on the end result, since the precision with which rockets have been used to inject

satellites into orbit suggests that aiming is quite accurate. The data are from *Economic Viability after Thermonuclear War: The Limits of Feasible Production*, memorandum prepared for U.S. Air Force Project Rand by Rand Corporation, Santa Monica, Calif., 1963.

8. It is perhaps worth mentioning that the Hudson Institute has carried out a considerable number of studies for the Department of Defense on civil defense problems and that the director of the Institute, Herman Kahn, is widely known for the view that in a nuclear war, properly organized and properly combined with civil defense, the nation could survive and even emerge as the "victor." It is relevant to note here that Mr. Kahn participated in a study program on civil defense sponsored by the National Academy of Sciences. A summary of the study group's report has been published as: *Project Harbor: A Report Prepared by the Office of Civil Defense and by a Summer Study Group* (Washington, D.C.: National Academy of Sciences, National Research Council, 1963). The full Project Harbor report is available on loan from the U.S. Army Library, Washington, D.C., This report also concluded that civil defense could be effective. Although the documents of the Project Harbor report make extensive reference to Hudson Institute studies that support Mr. Kahn's view, there is no comparable discussion, in these documents, of the Hudson Institute's extensive studies on the ecological effects of nuclear war.

The scenarios referred to on this and succeeding pages of this chapter are contained in a report published by the Hudson Institute: *Special Aspects of Environment Resulting from Various Kinds of Nuclear Wars*, Part II, H1303-RR, January 8, 1964 (Harmon-on-Hudson, N.Y.: Hudson Institute). A more recent report on these problems is Robert U. Ayres, *Environmental Effects of Nuclear Weapons*, Vols. 1, 2, and 3 (Harmon-on-Hudson, N.Y.: Hudson Institute, 1965).

9. "Post-Attack Farm Problems." Prepared for the U.S. Office of Civil and Defense Mobilization by Stanford Research Institute, Menlo Park, California, December 1960.

10. "Ecological Problems and Post-War Recovery: A Preliminary Survey from Civil Defense Viewpoint." Report prepared for U.S. Air Force Project Rand by Rand Corporation, Santa Monica, California, August 1961. For a detailed discussion of the ecological effects of nuclear war, see *Scientist and Citizen*, September-October 1963.

11. "Some Moral and Technical Consequences of Automation *(loc. cit.)* .

6: The Scientist and The Citizen

1. *Background Material for the Development of Radiation Protection Standards.* Staff Report of the Federal Radiation Council (Washington, D.C.: Federal Radiation Council Staff Report No. 1, May 13, 1960). For a detailed discussion of the illuminating history of the iodine-131 problem see *Scientist and Citizen*, September 1962 and September-October 1964.

2. For comprehensive expressions of the view that scientists ought to play a role in directing the power of modern technology, especially in connection with the issues of nuclear war, it is useful to read Eugene Rabinowitch, *The Dawn of a New Age: Reflections on Science and Human Affairs* (Chicago: University of Chicago Press, 1963) , and Leo Szilard, *The Voice of the Dolphins* (New York: Simon and Schuster, 1961).

3. *A Review of Space Research, op. cit.*

4. *Hearings before the Senate Committee on Aeronautical and Space Sciences, op. cit.*, p. 113.

5. "Science and Human Welfare," *Science* (July 8, 1960).

6. Readers can learn more about the work of the St. Louis Committee for Nuclear Information and its

publication, *Scientist and Citizen*, by communicating with CNI at 5144 Delmar Boulevard, St. Louis, Missouri 63108. Those interested in establishing contact with scientists in their own communities who are in a position to help them learn about problems of environmental pollution should write to Scientsts' Institute for Publication Information, 30 East 68th Street, New York, N.Y. 10021, which coordinates the work of some twenty scientists' information groups throughout the United States.

7. H. A. Knapp, "Iodine-131 in Fresh Milk and Human Thyroids Following a Single Deposition of Nuclear Test Fallout" (Washington, D.C.: AEC, 1963) , TID-19266.

8. The CNI testimony which recommended a medical survey of children exposed to radiation from the Nevada test site area is published by the Joint Committee on Atomic Energy in *Fallout, Radiation Standards, and Countermeasures.* Hearing before the Subcommittee on Research, Development, and Radiation. Part II (Washington, D.C.: Government Printing Office, 1963) , pp. 601-660.

9. The complete account, called "Nuclear War in St. Louis: One Year Later" was published in *Nuclear Information [Scientist and Citizen]*, September 1959, and in *Saturday Review*, November 28, 1959.

10. *Scientist and Citizen*, June 1961.

11. Dr. Pruitt's remarks are quoted from a presentation at the National Conference for Scientific Information held by the Scientists' Institute for Public Information, February 16, 1963.

12. Detailed reports of the work of the Baby Tooth Survey can be found in the following issues of *Scientist and Citizen:* March 1959; November 1961; March-April 1963; September-October 1964.

7: *To Survive on the Earth*

1. See "The Supersonic Transport," by Kurt H. Hohenemser, in *Scientist and Citizen*, April 1966.

2. For an illuminating discussion of the significance of the technological failure of the auto industry, see the article "The Murderous Motor Car" by Lewis Mumford in the *New York Review of Books* for April 28, 1966.

3. For a discussion of the urgency of transforming urban waste-removal systems, see *Waste Management and Control*, a report to the Federal Council for Science and Technology by the Committee on Pollution (Washington, D.C.: publication 1400 of the National Research Council, National Academy of Sciences, 1966). Among other things the report points out that at the present rate of accumulation of pollutants, essentially all of the surface waters of the United States will become so contaminated as to lose their biological capability for purification within the next twenty years.

4. Although it is apparent that runoff of fertilizers from farm land makes a growing contribution to the pollution of surface water by nitrate and phosphate, especially the former, detailed information about this problem is surprisingly scarce. As in the case of all pollution problems, little can be done about it until we really know the sources of specific pollutants. There is an urgent need for detailed study of agricultural fertilizer practices and their impact on the pollution of surface water.

Index

173

A SAND COUNTY ALMANAC

ALDO LEOPOLD

"There are some who can live without wild things, and some who cannot. These essays are the delights and dilemmas of one who cannot."

—Aldo Leopold

"We can place this book on the shelf that holds the writings of Thoreau and John Muir."

—*San Francisco Chronicle*